The Impact of the
Geosciences on Critical
Energy Resources

AAAS Selected Symposia Series

Published by Westview Press
5500 Central Avenue, Boulder, Colorado

for the

American Association for the Advancement of Science
1776 Massachusetts Ave., N.W., Washington, D.C.

The Impact of the Geosciences on Critical Energy Resources

Edited by
Creighton A. Burk and Charles L. Drake

AAAS Selected Symposium 21

AAAS Selected Symposia Series

Published in 1978 in the United States of America by
   Westview Press, Inc.
   5500 Central Avenue
   Boulder, Colorado 80301
   Frederick A. Praeger, Publisher

Library of Congress Catalog Card Number: 78-60678
ISBN: 0-89158-293-2

Printed and bound in the United States of America

About the Book

This volume focuses on the research achievements of the earth sciences in developing the nation's energy resources and on the efforts that still must be made toward solving current and future problems. Contributors point out that efficient exploration for energy resources, evaluation and development of these resources, and effective control of associated environmental factors depend largely on the basic concepts and knowledge developed within the geosciences.

About the Series

The *AAAS Selected Symposia Series* was begun in 1977 to provide a means for more permanently recording and more widely disseminating some of the valuable material which is discussed at the AAAS Annual National Meetings. The volumes in this *Series* are based on symposia held at the Meetings which address topics of current and continuing significance, both within and among the sciences, and in the areas in which science and technology impact on public policy. The *Series* format is designed to provide for rapid dissemination of information, so the papers are not typeset but are reproduced directly from the camera-copy submitted by the authors, without copy editing. The papers are organized and edited by the symposium arrangers who then become the editors of the various volumes. Most papers published in this *Series* are original contributions which have not been previously published, although in some cases additional papers from other sources have been added by an editor to provide a more comprehensive view of a particular topic. Symposia may be reports of new research or reviews of established work, particularly work of an interdisciplinary nature, since the AAAS Annual Meetings typically embrace the full range of the sciences and their societal implications.

WILLIAM D. CAREY
Executive Officer
American Association for
the Advancement of Science

Contents

List of Figures

List of Tables

About the Editors and Authors

Creighton A. Burk, director of the Marine Science Institute and professor and chairman of the Department of Marine Studies at the University of Texas at Austin, was previously chief geologist with Mobil Oil Corporation and manager of its Regional Geology Group. His work has concentrated on worldwide tectonics and the geology and geophysics of continental margins, and his extensive publications in these fields include The Geology of Continental Margins *(edited with C.L. Drake, Springer-Verlag, 1974). Dr. Burk is U.S. delegate to the US-USSR Protocol in Oceanography, chairman of the Marine Geology Committee of the American Association of Petroleum Geologists, and a fellow of the Geological Society of America. He was U.S. delegate and chairman of the panel discussion on "Global Tectonics and Petroleum Occurrences" for the 1975 World Petroleum Congress and a member of the American Geological Institute Committee on Geoscience and Public Policy.*

Charles L. Drake, chairman of the Department of Geological Sciences at Dartmouth College, specializes in marine geology and geophysics, tectonics, structural geology and seismology. He is past-president of the Geological Society of America, president of the Inter-Union Committee on Geodynamics, International Council of Scientific Unions, and chairman of the Committee on Geodynamics of the National Academy of Sciences. Dr. Drake's publications include Geodynamics: Progress and Prospects *(American Geophysical Union, 1976).*

Allen F. Agnew is a senior specialist in environmental policy (mining and mineral resources) with the Congressional Research Service of the Library of Congress. He is chairman of the Committee on Environment and Public Policy of the Geological Society of America, and a former chairman of the Ad Hoc Committee on Public Statements of the American Geological Institute. His many publications cover a wide range of subjects, including general geology, mineral deposits, water resources, and related public policy topics.

William L. Fisher, director of the Bureau of Economic Geology and chairman of the Council on Energy Resources at the University of Texas at Austin, is former Assistant Secretary of the Interior for Energy and Minerals. He is an American Association of Petroleum Geologists Distinguished Lecturer and the author of numerous publications on the geology of energy, mineral resources and resource policy.

Peter T. Flawn, professor of Geological Sciences and Public Affairs at the University of Texas at Austin, is president of the Geological Society of America (1978) and a member of the National Academy of Engineering. His work has focused on economic geology, environmental geology, and public policy, and his publications include **Mineral Resources** *(Rand McNally, 1966) and* **Environmental Geology** *(Harper and Row, 1970).*

John D. Moody is an international energy consultant in New York. He is former president of the American Association of Petroleum Geologists.

Leon T. Silver, professor of geology at the California Institute of Technology, is vice-president of the Geological Society of America (1978). He is a member of the National Academy of Sciences, chairman of its CONAES Uranium Resource Subpanel, and former chairman of its Board on Mineral and Energy Resources Workshop on Uranium. His numerous publications concern the geology and geochemistry of uranium and the applications of uranium-thorium-lead isotope systems to geochronology, crustal evolution, ore deposits and geological history. Professor Silver has received awards from NASA, the American Institute of Professional Geologists, and the Geological Society of America.

Jack A. Simon, chief of the Illinois State Geological Survey, is also a professor in the Department of Metallurgy and Mining Engineering at the University of Illinois. His work has focused on coal geology, and he has published extensively on many aspects of this topic. He is an honorary life member of the Illinois Mining Institute, a fellow of the Geological Society of America and of the American Association for the Advancement of Science, and a recipient of the Gilbert H. Cady Award from the Geological Society of America for his contributions to coal geology.

M. Gordon Wolman, professor of geography and chairman of the Department of Geography and Environmental Engineering at *The Johns Hopkins University, specializes in geomorphology and environmental geology. He is a council member of the American Association for the Advancement of Science and of the American Geophysical Union; a member of the National Academy of Sciences/National Research Council Committee on Minerals and the Environment, and past chairman of the NAS/ NRC Committee on National Water Quality Policy (1973-76). He has published on the subjects of river processes, sedimentation, and environmental policy, including* **Fluvial Processes in Geomorphology** *(with L.B. Leopold and J.P. Miller, 1964).*

Overview

Charles L. Drake and Creighton A. Burk

At the symposium marking the 30th anniversary of the Office of Naval Research, Lewis Thomas of the Sloan-Kettering Institute addressed the subject of biomedical research and biomedical technology.[1] He observed that in the case of infectious diseases, the causes of which are understood through 100 years of biomedical research, control is simple and inexpensive. Smallpox, polio, typhus or typhoid all can be prevented through a simple vaccination. Then there are other diseases -- heart trouble, hypertension, cirrhosis, arthritis, stroke or cancer that present biological problems for which the fundamental scientific knowledge is not yet available. In the absence of this knowledge; money and technology are thrown at the problems, perhaps easing the pain, but not affecting a cure.

In 1953 our economy developed a serious heart pain whose apparent cause was a large and seemingly arbitrary increase in the cost of hydrocarbons. This great pain was only a symptom of the real disease, but we will respond, as we do for human cancer, by trying to ease the pain through the application of money and technology. Our energy medicare plan will include regulation which inhibits the search for alternatives by creating artificial imbalances among the costs of existing energy sources. Perhaps some of this must be done in order to provide short-term relief, but what about the long-term -- can we affect a cure? What, for that matter, is the basic nature of disease?

In the United States, and in the era of Cheap energy, we have developed the habit of using energy in a profligate manner. Costs were so low as compared to income that we used it almost as if it were free. We do the same with water, an even more important commodity that also promises to be in short supply with time. We have rewarded heavy users of both energy and water by providing it at rates far lower than those

that the small user must pay. We encourage use of natural
gas through artificially depressed prices relative to alternate energy sources, and through liberal environmental constraints since it is the cleanest of the fossil fuels. We
are so emotionally torn by the perceived hazards of nuclear
fuel sources that we almost forget that they are finite. We
are told that we can always fall back on coal in a pinch, but
the recent coal mine strike illustrates the dangers of being
dependent upon a single energy source.

How can the geosciences contribute in this area? The
most obvious way is in accurately assessing the energy resources that are technically, politically and economically
available. These assessments vary widely. In this volume
Simon notes that estimates of even "recoverable reserves" are
still too inexact to have any meaning; Silver cites the unknowns concerning uranium resources and the need for high-
quality exploration programs; Moody notes the large variations that have existed in estimations of hydrocarbon resources and reserves; Fisher cites variations of estimates of
the energy resources in the geopressured zones of the Gulf
Coast as varying by a factor of 500. Clearly we must do better, given the scale and degree of interdependence within our
society, and equally we cannot do better without good data,
intelligently interpreted.

In addition to determining the magnitude of these resources, we need to know where they are. This is a task with
many facets. We need to know the history of development of
the Earth's crust in which the resources are found. We need
to know the processes by which the elements in question are
concentrated, moved to their present locations and trapped.
We need to know where they are trapped and in what quantities. We need to know the economic, energy, political and
environmental costs of extracting them or disposing of their
residue. In each of these areas there are unknowns that can
be lessened through knowledge acquired from fundamental geoscience research.

How are we progressing in this regard? Flawn notes that
university research and teaching programs in the area of
earth resources is at a surprisingly low ebb. The focus of
Federal support to the geosciences in universities has been
on basic research, as it should be, but little of this has
gone into basic resource research. This may be because the
mission of the National Science Foundation, the primary supporting agency, is to ensure that there is a viable scientific community rather than being directed towards practical
problems. It may be that the scientific community has developed the Nobel Prize syndrome and focusses on the heroic

scientific breakthrough rather than the science more closely related to the problems of our society. The mission-oriented agencies such as Interior or Energy do not have a good track record at present in support of resource research in the geosciences in the universities, but perhaps there is some hope for the future. Industry support for the geosciences in the universities has never attained a level commensurate with either industry or national needs. Industry, the Federal agencies and the universities should enter into discussions of the research needs in the geosciences, if this can be done in these days of confrontation politics.

Communication among the actors in the energy passion play are not all they should be. Agnew notes that about fifteen Federal agencies deal with geoscience issues in one way or another. The rapport between these agencies and state geologists is improving, but has a long way to go. Communication with university geoscientists is quite variable, perhaps because science research at the universities is at a low ebb, or perhaps because there is a lack of knowledge on how to communicate. Communication between the agencies and industry is inhibited by concerns about conflicts of interest. It is not clear that the advisory mechanisms that are available to the Congress or the Executive branch of the Federal Government are always used to the best advantage.

Geoscience research will provide no panaceas for solving the energy problems of the country. However, absence of the knowledge that would result from such research will certainly inhibit the finding of solutions. The opportunities and challenges for geoscientists in government, in industry, and in the universities are manifold. Let us hope that they and their sources of support will be responsive to these challenges and that the information resource base will begin to reflect the real natural resource base and can provide more than fantasy to the policy resource base. This may help to make it possible to think in terms of a cure for our energy problems rather than merely an alleviation of pain.

1. Thomas, Lewis, 1977: Frontiers in Biomedicine and Biomedical Technology, in Science and the Future Navy, National Academy of Sciences, p. 53-61.

1

Federal Objectives and Organizations Relative to the Geosciences

Allen F. Agnew

The Geosciences and Energy

Today, virtually everything we use and do somehow depends on energy--for example, our transportation and the manufactured products we use. The tremendous rise in U.S. energy consumption in the past decade or so has been due to an increase in per-capita use rather than to an increase in our population. Furthermore, the U.S. is more and more in competition with other countries of the world for energy resources; --this situation will be exacerbated in the future, as the less-developed countries attempt to satisfy their desire for higher living standards.

Energy resources on which we depend today are the energy fuels--oil, gas, coal, and uranium--plus hydroelectric power. In his book on Geology, Resources, and Society, H.W. Menard pointed out that the U.S. accounted for 35 percent of the total world energy consumption in 1967; this ranged all the way from only 20-25 percent of world consumption for coal, hydroelectric, and nuclear, up to a high of 64 percent for gas. (see Table 1).

Some of these energy sources are more limited than others, and some of the less developed and more exotic sources (geothermal, solar) will require substantial investment in development technology before they can be considered feasible for even long-range planning. Existing sources of energy will require additional investment in exploration and development so that their full potential can be tapped; thus our attention should continue to be placed on oil and gas, coal, oil shale, and nuclear, while at the same time we look to other sources in the future.

All sources of energy that are presently in use lead to pollution--ranging from relatively little to great in amount--

Table 1 - World energy consumption, 1967, by the U.S. and the rest of the world, in percent of kilowatt-hours

	Oil	Gas	Coal	Nuclear	Hydro-electric	Over-all
U.S.	35.7	64.1	20.2	25*	21.8	34.8
Other Developed Countries	50.9	30.5	60.9	75*	63.4	51.1
Less Developed Countries	13.5	5.4	18.7	0	14.9	14.0
Total	100.1	100.1	99.8	100*	100.1	99.9

* Data for USSR not available
  Source: Modified from Menara, Henry W. Geology, Resources and Society. W. H. Freeman and Company, 1974 (Table 18.1)

and all of them cause at least local environmental change. Thus, air and water pollution, environmental insults, and unhealthy effects on living things are the negative aspects of our massive hunger for energy.

All of these sources and effects are the result of geoscience processes or modifications thereof by people--but they are also amenable to the healing effects of geological processes. Thus, just as there has been a substantial challenge to geoscientists in the past century to devise means of locating, extracting, and harnessing these resources, there is an even larger challenge today and in the future for us to continue those resource-supply efforts while at the same time applying geoscience knowledge and ingenuity to the solution of environmental problems that have resulted in the past and will continue to occur in the future. The geosciences are therefore deeply involved in virtually all aspects of energy matters, because they affect not only us individuals, but also the environment which provides us with the standard of living (or quality of life) that we hold dear. Let us now turn to see how the federal government interacts with the geosciences in these energy matters.

Objectives and Agencies

Objectives

Politics and administration must be kept separate, and efficiency is the overriding goal of organization and administration. At least, that's what Harold Seidman thought, before he began his 25 years of observing the Federal scene "through a particular window in the Bureau of the Budget." (Seidman, 1970, p. vii) He became increasingly aware, however, that few issues are not political ones, and that an overriding concern is the balance of power among several contending forces, as follows:

--the President of the U.S.;
--Congressional committees;
--the bureaucracy;
--State and local governments; and
--organized groups in the private community which are
    affected in one way or another by Federal policies and
    programs. (Ibid, p. vii-viii)

Reorganization has become almost a religion in Washington, D. C.--and reorganization is deemed synonymous with reform, and reform with progress. So said Seidman, and he added that

> "periodic reorganizations are prescribed if
> for no other purpose than to purify the bureau-
> cratic blood and to prevent stagnation." (<u>Ibid</u>, p. 3)

The devils to be exorcised, he continued, are overlapping and
duplication on the one hand, and confused or broken lines of
authority and responsibility on the other. The twin bless-
ings of economy and efficiency can be obtained only by strict
adherence to sound principles of executive branch organiza-
tion--or so the fundamentalist dogma holds.[1] (<u>Ibid</u>, p. 4)

Public administration, says Seidman, is viewed as being
concerned almost exclusively with the executive branch, with
only grudging recognition given to the roles of the legislative and judicial branches. (<u>Ibid</u>, p. 6) This preoccupation
is coupled with a distrust of politics and politicians, who
it perceives as the natural enemies of efficiency.

Since World War II, however, public-administration dissenters from this dogma have been increasingly vocal in their
disenchantment with the orthodox. A few of these dissenters
have suggested alternatives, such as "adaptive, rapidly
changing temporary systems" that are organized around problems needing solution. (Bennis, 1966, p. 12) However, as
Seidman comments (1970, p. 8), individual members of Congress
can relate to dogma and its set of guidelines--which are simple, readily understood, and comprehensive[2]--and thus the
Congress has insisted on more, not less orthodoxy.

Although nearly every President of the U.S. in the twen-
tieth century through Lyndon B. Johnson defended reorganization as a means of reducing expenditures, the overwhelming
evidence supports the Franklin D. Roosevelt view that reor-
ganizations do not save money. (<u>Ibid</u>, p. 12) The reason for
reorganization, said FDR, is good management. (Polenberg,
1966, p. 8)

Unfortunately, reorganization has placed more emphasis
on form than substance--thus showing an incomplete understanding of our constitutional systems, of institutional behavior, and of the tactical and strategic uses of organiza-
tional structure as an instrument of politics, position, and
power.

Actually, the structure of the executive branch is a
microcosm of our pluralistic society--reflecting its conflicting values and competing forces. (Seidman, 1970, p. 13)
The struggle for power and position among those entities that
have become increasingly dependent, since World War II, on
the federal government, and its financial aid has contri-

buted to a fragmentation of the structure of the executive branch and to the proliferation of its categorical programs.[3]

With the preceding discussion as background, let us now examine briefly the executive agencies that deal with the geosciences--one of which, the Department of Energy, has recently been born out of just such a reorganization.

The Geosciences and the Agencies

From the foregoing discussion we can see how difficult it would be to compile an exhaustive listing of executive agencies that deal with the geosciences. And it is difficult, as I learned. Part of the trouble lies in the fact that the geosciences touch so many facets of our total existence--not only the resources, services, and construction which form the basis for our society and its standard of living, but also the less-tangible humanistic elements of the quality of life that we enjoy.[4]

The other major reason for this difficulty is that most agencies are broad and general in thrust, and consist of a number of subelements that range greatly in cohesiveness and interrelatedness. For example, the Department of the Interior contains at least six entities that deal with the geosciences in varying degrees, as follows:

--Geological Survey;
--Bureau of Mines;
--Bureau of Land Management;
--Bureau of Indian Affairs;
--Fish and Wildlife Service; and
--National Park Service.

One way to consider the scope of the involvement of the geosciences in U.S. living today is to examine the major programmatic areas of the U.S. Geological Survey, as shown in Table 2. A subjective scanning of executive agencies shows that nearly all of the major units possess one or more subdivisions that have at least a minor interest in, or dependence on, the geosciences [5] (Table 3); those having a substantial programmatic effort in the geosciences are indicated by asterisks.

Legislation

We must not lose sight of the fact that another branch of the Federal government plays an exceedingly important role in establishing and otherwise influencing policy, the Congress--by the legislation that it enacts (or does not en-

Table 2 - Broad Programmatic Areas of the U.S. Geological
Survey

Energy   -- Outer continental shelf oil and gas
                        -- Onshore oil and gas
                        -- Coal
                        -- Uranium, geothermal, and oil shale

Minerals   -- Metalic, nonmetallic, and ocean minerals

Hazards   -- Earthquakes, volcanoes, and landslides

Fundamental geologic knowledge (includes geophysics and
                        geochemistry)

Water   -- data networks and special studies (both
                        fundamental and applied)

Land-resource decision products
* Topographic mapping
* Earth resources observation systems (remote sensing)
* Provides services used by all foregoing programmatic areas
  Source: U.S. Geological Survey

Table 3 - Federal Executive Entities Related to the
Geosciences, Even Peripherally (includes
quasi-Federal entities)

Executive Office of the President

   White House
   Office of Management and Budget
   Council of Economic Advisors
   National Security Council
   Domestic Council
   Council on Environmental Quality
   Council on International Economic Policy
   Energy Resources Council
   Office of Science and Technology Policy

Departments

   **Agriculture
   **Commerce
   **Defense
   **Energy
   **Health, Education, and Welfare
      Housing and Urban Development
   **Interior
    *State
    *Transportation
      Treasury

Agencies

    *Appalachian Regional Commission
   **Environmental Protection Agency
   **National Aeronautic and Space Administration
   **National Science Foundation
      Overseas Private Investment Corporation
   **Tennessee Valley Authority
   **National Academy of Science/National Academy of
      Engineering
    *Smithsonian Institution

Selected Multilateral International Organizations
Selected Bilateral International Organizations

    * Have substantial programs dealing with the geosciences
   ** Principal ones
      Source: Adapted from Government Manual, 1976-1977.

Table 4 - Congressional Committees with which the USGS Had Substantial Interaction in 1977

Senate

Appropriations/Interior

Commerce, Science & Transportation/Science, Technology & Space

Energy & Natural Resources/Energy R & D
Energy & Natural Resources/Public Lands & Resources

Environment & Public Works/Water Resources

House

Appropriations/Interior

Government Operations/Environment, Energy & Natural Resources
Government Operations/Legislation & National Security

Interior & Insular Affairs/Energy & Environment
Interior & Insular Affairs/General Oversight & Alaska Lands
Interior & Insular Affairs/Indian Affairs & Public Lands
Interior & Insular Affairs/Mines & Mining
Interior & Insular Affairs/National Parks & Insular Affairs

Interstate & Foreign Commerce/Energy & Power
Interstate & Foreign Commerce/Oversight & Investigations

Merchant Marine & Fisheries/Oceanography

Science & Technology/Advanced Energy Technology & Energy Conservation RD&D
Science & Technology/Environment
Science & Technology/Fossil & Nuclear Energy RD&D
Science & Technology/Space Science & Applications

Ad Hoc Select Committee on OCS

Joint

Economics

Source: U.S. Geological Survey

act) which authorizes Executive Branch programs. The Congress also appropriates the funds that enable the executive branch to carry out these programs. Thus the objectives of the Federal government concerning the geosciences (or any other subject, for that matter) are defined and implemented by a combination of interactions on the playing field of policy-making between the Executive Branch and the Congress, sometimes--it has been said--in an adversarial manner.

However, as the late Senator Hubert Humphrey said, the Congress should not be viewed as "a battlefield for blind armies that clash by night. This is a place where national objectives are sought--where Presidential programs are reviewed--endlessly debated and implemented." (Senator Hubert H. Humphrey, in Tacheron and Udall, 1970, p.9) Senator Humphrey commented in his 1965 Commencement Speech at Syracuse University that two lessons he learned from his collegial experience in the Congress deal with: (1) the creative and constructive dimension of the process of compromise and (2) the importance of the Congressional role of responsible surveillance of executive programs. (Ibid, p.7-9)

To show the breadth of Congressional committees that are involved with geoscience-related issues, it is instructive to examine the subcommittees that the U.S. Geological Survey deal with (See Table 4).

In keeping with the theme of this Symposium, "The Impact of the Geosciences on Critical Energy Resources," let us now restrict our further consideration to those aspects of the geosciences that deal with energy resources. Such a narrowing of focus will not significantly reduce the number of total legislative issues to consider, when we examine those that received some degree of action during the first session of the present Congress.

Table 5 shows those geoscience-related energy issues which went the whole route and became public laws in 1977.

Table 6 shows those issues for which the legislative process has not yet been completed; they have been carried over into the second session of the Congress, which began just three weeks ago--and hearings and markup sessions are under way for some, a few of which will likely be enacted. Others, as is the history of the legislative process, will die with the end of this Congress next fall. However, because they represent current or recurring issues of national concern, many of them will be reintroduced in the 96th Congress, which will convene in January, 1979.

Table 5 - Geoscience-related Energy Issues Enacted by the
95th Congress, First Session (alphabetically)*

Alaska Native Claims Settlement Act --P.L. 95-178
Alaska Natural Gas Transportation System --P.L. 95-158

Armed Forces: Defense Production Act --P.L. 95-37

Atomic Energy: Nuclear Regulatory Commission --P.L. 95-209

Environment: Clean Air Act Amendments --P.L. 95-95
Environment: Environmental Protection Agency, authorization
              --P.L. 95-155

Fuels & Energy: Coal Conversion --P.L. 95-70
Fuels & Energy: Energy Department, establish --P.L. 95-91
Fuels & Energy: ERDA R & D authorization --P.L. 95-39
Fuels & Energy: Oil & Gas Lease, Deep Drilling, extend
                 --P.L. 95-77

Mining, Surface, Control and Reclamation Act --P.L. 95-87

National Science Foundation, authorization --P.L. 95-99

Water: Clean Water Pollution Control Act --P.L. 95-217

---

* plus appropriations authorizations for numerous executive
  departments and agencies
  Source: Congressional Record, January 4, 1978, Index of
           Legislative Business Transacted, p. D1723-D1731

But national objectives are not determined solely by people in Washington, D.C.,for the Congress regularly seeks advice from experts and concerned lay people who reside in various parts of the nation. And the Executive Departments make use of advisory groups and public meetings to elicit suggestions as to views on national issues as seen by those with regional or State or local perspective. For example, a liaison committee of the Association of American State Geologists meets regularly with the USGS and a number of other Federal agencies regarding geoscience issues of national scope, thereby bringing the talents of a number of geoscience experts who are scattered around the country to bear on such matters.

Professional geoscience associations such as the American Association of Petroleum Geologists, the Society of Exploration Geophysicists, the American Geophysical Union, and the Association of Professional Geological Scientists, likewise provide objective, expert testimony and people at the national level--both to the Executive Branch and to the Congress--and thereby aid in the establishment and implementation of national policy.

As one who has been deeply immersed in the national legislative scene during the past four years, I am impressed with the breadth of applicability of the geosciences, and their potential value (if fully harnessed) to help solve the wide range of issues of national importance that depend upon knowledge and application of the geosciences. The six speakers who follow me will present graphic evidence of the many ways in which the geosciences can influence the critical choices regarding energy which this nation must make, both today and in the future.

Our job is mainly one of communication--and it is the task of all geoscientists to see that we perform that job both continuously and well.

Table 6 - Geoscience-related Energy Issues Considered but
       not Completed by the 95th Congress First Session
                    (alphabetically)*

Alaska Oil, Transportation (S.1868, HR 9203)

Atomic Energy: Nuclear Energy Disposal (HR. 6181)

Environment: Alaska National Interest Lands, Conservation
              (HR 39)
Environment: Environmental Information System (S657)
Environment: Toxic Substances Control Act (S1069, 1531)
Environment: Toxic Substances
Federal Power Commission, authorization (S1535)

Fuels & Energy: Coal Slurry Pipeline Act of 1977 (HR 1609)
Fuels & Energy: Coal Utilization (HR 5146, 8444, S977,
                 S. Res 228)
Fuels & Energy: Competition in Energy Industry (S1927,
                 HR 7816)
Fuels & Energy: Energy Conservation (S2057, HR 5037, 8444,
                 S. Res 257)
Fuels & Energy: ERDA Civilian Nuclear Applications (S1811,
                 S. Con. Res 55, S. Res 216, HRes 657,
                 Con Res 384)
Fuels & Energy: ERDA Nonnuclear Authorization (S1340,
                 HRes 916)
Fuels & Energy: ERDA Synthetic Fuels Loan Programs (S37)
Fuels & Energy: Fossil R & D Program
Fuels & Energy: Helium, conserve (HRes 91)

Table 6 (continued)

Fuels & Energy:	Natural Gas Pricing (HR 5289, 8444, S2104, 256)
Fuels & Energy:	Oil Pricing, Crude, New and Old
Fuels & Energy:	Oil Prices/OPEC
Fuels & Energy:	Oil Shale Technology (S 419)
Fuels & Energy:	OCS Resource Management (S9, SRes 204, HR 1614)
Fuels & Energy:	Petroleum Marketing Practices (HR 130)
Fuels & Energy:	Pipeline, Destruction of, Interstate and Trans-Alaska (S1502)
Fuels & Energy:	President's Proposed Financial Reporting System by Energy Companies
Fuels & Energy:	Public Energy Competition Act (HR 7816)
Fuels & Energy:	Strategic Petroleum Reserve, oversight
Fuels & Energy:	Uranium Mill tailings, Cleanup (S266, SRes 129)
Fuels & Energy:	Wyoming Oil & Gas Leases (S259)
Marine Matters:	Ocean Pollution research (S1617, HR 7878)
Mine Safety & Health Amendments (HR 4287, S717, HRes 682, S Con. Res 57)	
Parks & Recreation:	Alaska Lands (HR 6565, 1652, 5605)

* plus appropriations authorizations for several executive departments and agencies

Source: Same as Table 5

References

Bennis, Warren G., 1966, Changing Organizations: Essays on the Development and Evolution of Human Organization: New York, N.Y. McGraw Hill, 223 p.

Gulick, Luther, 1937, "Notes on the Theory of Organization," p. 1-45, in Papers on the Science of Administration; edited by Luther Gulick and L. Urwick; Inst. of Public Adm., Columbia Univ. 195 p.

Landsberg, Hans H., 1964, Natural Resources for U.S. Growth: to the Year 2000: Baltimore Md., John Hopkins Press, 260 p.

Menard, Henry W., 1974, Geology, Resources and Society, San Francisco, Calif., W. H. Freeman & Co., 621 p.

Polenberg, Richard, 1966, Reorganizing Roosevelt's Government; Harvard Univ. Press., 275 p.

Seidman, Harold, 1970, Politics, Position, and Power: New York, N.Y., Oxford University Press, 311 p.

Senate Committee on Government Operations, 1966, Hearings on "Federal Role in Urban Affairs," Aug. 15-19, 22-26, 29-31, Sept. 1, Nov. 29-30, Dec. 2, 5-9, 12-15, 1966: U.S. Cong., 89th, 2nd Sess., 3034 p. plus 104-p. Appendix, Dec. 30.

Government Manuel, 1976-77, United States: Washington, D.C., U.S. Government Printing Office, 871 p.

Notes

1. Unfortunately, the dogmatists have overlooked the important caveats and qualifications of Luther Gulick (1937, p. 3), who emphasized the futility of seeking a single most effective system of departmentalism and emphasized the need to recognize that, "organization is a living dynamic entity", that there are limitations to command, and that leadership plays a significant role.

2. In contrast to the clarity of the established dogma, Seidman said, proponents of newer orthodoxies have tended to write for each other in an arcane language that is untelligible to the lay public. (It seems to me that we geoscientists-- and other scientists, too, for that matter--have also heard such criticism directed at ourselves, and more than once, unfortunately.)

3. By narrowing their constituencies, agencies become more susceptible to domination by their clientele groups and congressional committees. Efforts to narrow their constituencies have been accompanied by demands for independent status within the departmental structure. In fact, the fundamental question was posed by Senator Robert Kennedy in 1967, when he asked,
> "Do the agencies of government have the will and determination and ability to form and carry out programs which cut across departmental lines, which are tailored to no administrative convenience but that overriding need to get things done." (Senate Com. Govt. Ops, 1967, p. 40)

4. For example, more than a decade ago it was estimated that our resources would be required to supply for the kinds of output in the amounts or proportions indicated in Table 7. Thus we see that the value of services is twice as great as that of industrial production, which is twice as great as that of construction; the relative value of agricultural production was only one third as great as that of construction in 1960, and its relative proportion is projected to decrease to only one-sixth by the year 2000.

Table 7 -- Real and Projected Output Distribution
(in billions of dollars, with 1957=100)

	1960	%	1980	%	2000	%
Services	189	50.5	415	50.5	909	50.5
Industrial Production	108	29	249	30	564	31.5
Construction	57	15	130	15	281	15.5
Agricultural Production	21	5.5	29	3.5	38	2
Total	375	100.0	823	100.0	1792	99.5

Source: Landsberg, Hans H. Natural Resources for U.S. Growth:
to the year 2000. Copyright © The Johns Hopkins
University Press for Resources for the Future, 1964
(Figure 1)

5. This may take the form of a fully identifiable bureau such as the U.S. Geological Survey (Department of the
Interior); lesser offices such as the National Ocean Survey
and National Geodetic Survey of the National Oceanic and
Atmospheric Administration (Department of Commerce); or only
program elements.

2

The Impact of University Geoscience Programs on Critical Energy Resources

Peter T. Flawn

The term <u>resource base</u> (as defined precisely by Shurr in 1960) is commonly used to express the total amount of a mineral commodity that may be available. It includes what has been, is, or may be within a specified geographic region -- it includes the discovered (and already produced), the yet-to-be discovered, the economically producible, and the amounts of the commodity that might be produced under various sets of economic conditions and operational technologies. It is our endowment. Unfortunately, the term is used not wisely but too casually. And thus calculations are made to tell us the resource base for oil in various size barrels or tons, natural gas in cubic feet or meters, and coal in short, long, or metric tons and uranium in "forward costs". And with these figures, it is easy for almost anyone to look up current production and calculate how many years' supply we, any other nation, or the world has, and to recommend policies accordingly.

After some decades of playing around with resource bases and building models out of them, it should be clear to all that despite the interesting fantasies that have been woven, knowing the U. S. resource base for, let's say, coal or uranium, or oil or gas does not seem to tell us much about how much we can produce next year or ten years from now. Knowing how much of a commodity you have, where it is, and what the quality is, simply is not enough if you are worried about ore in the bin or oil in the tank.

So--if you are worried about production or supply, there are other "resource bases" in addition to the resource base that expresses the physical existence of the commodity, including (1) the base of professional, technical and laboring minds and hands--the manpower resource base, (2) the financial resource base of investment and operating capital, (3) the technology resource base for finding, producing, con-verting, and (4) that all important <u>policy</u> <u>resource</u> <u>base</u> that makes it all possible—the one that changes the other resource base fantasies into reality. It is the policy resource base that makes it possible for a nation to produce a piece of its physical resource base and convert it into something useful—to make it truly a <u>resource</u>. I suppose it is only a matter of time until a model builder integrates the manpower resource base, the financial resource base, the technology resource base, and physical resource base to generate a producibility index. I am sure that we will have the model before we have a producing oil field on the Atlantic Outer-Continental Shelf. But the policy resource base, even cast into scenarios, presents the ultimate challenge to the model builder.

We are assembled to examine the impact of the geosciences on critical energy resources and I have an even more limited task—to assess the university's contributions to "the impact". It is immediately obvious that the geosciences impact on the physical resource base, the technology resource base, and the professional and financial resource base. I wish I could say that the geosciences have had even a modest impact on the policy resource base. However, despite the efforts of the very few geoscientists who have preached in the house of policy, the realities of nature and the distribution of mineral deposits have not been built into our policy resource base. I leave it to other participants in this symposium to develop this theme.

The university (1) supplies professional and technical manpower in geosciences and related engineering fields, (2) carries on fundamental and applied research in the geosciences and related fields, (3) carries out independent, integrative, objective, evaluative policy analysis, and (4) supplies, from time to time, professors to fill technical and policy positions in government. The American Enterprise Institute in a 1977 symposium (Professors, Politicians, and Public Policy) concluded that professors are probably the single most influential group in public policy in the United States (SCIENCE 14 August 1977, p. 742). However, there are all too few professors of geosciences in the increasing influential number of academics in government. University impacts on critical national problems are rarely visible, but in a technological, industrial society the university's human and research products provide a fundamental underpinning for the accomplishment of national purpose. If the flow of professionals and research from the universities into the national resource base diminished or ceased, our society would, in a decade, suffer from a seriously diminished capability to maintain itself.

This symposium is focused on geosciences, so I leave aside the university's role in engineering, physics, chemistry, economics, policy analysis, and the other disciplines and professions that deal with "critical energy resources."

University programs in geosciences suffer from weaknesses of many years standing. The geosciences span a wide range of related disciplines so that generalization is difficult and the program weaknesses are not evenly distributed. Those geoscience fields that impact directly on critical energy resources are (1) mining geology (economic geology), (2) petroleum geology, and (3) exploration geophysics. The market for graduates in these fields was very strong in 1977 and stronger in 1978. Recruiters are reported as having difficulty finding the kinds and number of graduates desired. However, mineral exploration has always been closely tied to the overall state of the economy. Exploration budgets are trimmed early in a down cycle and exploration geologists are out of work first. This lack of job security for explorationists is a matter of concern in university geoscience departments and has a negative effect both on industry-university relations and on student election of geoscience careers. A major employer's decision to trim or eliminate an exploration department "dumps" significant numbers of exploration geologists on the "market" and perturbs the normal flow of university graduates into the private sector.

In mining or economic geology, the opportunities have been so limited for so many years that there is a serious shortage of experienced mid-career scientists capable of applying the science to the solution of pressing mineral resource problems. Lack of opportunity in the 1960's forced career paths of young earth science professionals into other directions. The combination of training and experience required to produce a qualified mining geologist is such that geologists trained in other fields cannot readily apply themselves to mining geology problems without extensive retraining and apprenticeship. There is a good correlation between the number of prospects and mines studied and the effectiveness of the mining geologist. In mining geology there is a clear need for a better bridge between university programs and industry needs. Heavy emphasis in universities on theoretical aspects of ore genesis has been at the expense of field studies of mineralized districts.

Petroleum geology has fewer manpower problems than mining geology because of the existence in U. S. universities for 30 years of relatively large, healthy, well-attended general geological programs whose graduates, while not majoring in petroleum geology as such, can be productive in petroleum geology with a relatively short on-the-job training program. Students of stratigraphy, sedimentation, and structural geology, for example, are through these disciplines productive in the search for oil and gas. Of great current interest in oil and gas exploration are the fields of seismic stratigraphy, environmental geochemistry, and plate tectonics.

Exploration geophysics can likewise draw upon a relatively large pool of well-trained computer scientists, electrical engineers, and physicists who can, in pursuit of enlarged opportunity, be converted from other fields without excessively long lead times. Seismic exploration geophysics is today so automated little or no interpretation is done in the field. Data on tapes is fed into company centers where interpretation is carried out by sophisticated techniques at a relatively high level.

In exploration geophysics that utilizes electrical, magnetic, or gravity techniques there is not the same degree of automation. Field interpretation is an important capability. However, in 1978 there does not seem to be a manpower problem in this area. Both graduates and jobs are available. With these techniques, as for seismic techniques, there is a manpower flow from scientific and engineering disciplines other than traditional geosciences--to fill the need.

University research funding for geoscience programs in mining geology, petroleum geology and exploration geophysics is mostly in the category of NSF small grants to individual principal investigators. Sustained, continuing programs are rare. They are found mostly in those universities that house State Geological Surveys. The geoscience "research well" has been relatively dry since the days of ARPA's Vela Uniform project and AEC-funded university uranium resource programs. The logical federal agency to support university geoscience research, the United States Geological Survey, has traditionally been an in-house research organization. Over the years contracts and grants from the USGS to universities have been minimal.

University geoscience departments received considerable benefit from NASA funding for the Lunar Science Program. While the funds were not directed to mining or petroleum geology, or exploration geophysics, these programs received indirect benefits from rebuilt laboratories and new equipment.

What is new and encouraging is a number of university-established Energy Research Centers, some of which have already received substantial support from ERDA (now DOE). If DOE continues to support University Research it may be the beginning of a new era for university geoscience programs.

Another positive step came through the Surface Mining Control and Reclamation Act of 1977 which establishes State Mining and Mineral Resources and Research Institutes at a public college or university in a participating state. However, at the time of this writing, the centers have not been funded.

The most serious deficiency in university programs that impact on critical energy resources is in mineral technology. In 1969, the National Research Council reviewed this field and published a report entitled "Mineral Sciences and Technology--Needs, Challenges, and Opportunities". The report concluded (p. 1) that "Support of university research and graduate programs in mineral science and engineering is far below the level needed if the United States is to maintain a pre-eminent position in Mineral Technology." In the eight years since that report was published the situation has deteriorated. Programs have been abolished; enrollments have declined; lack of research funding has discouraged faculty. The existing cadre of experienced well-trained faculty were educated before 1950 and will be retiring over the next 10 or 15 years without younger faculty to take their places.

The establishment of Mining and Mineral Resources and Research Institutes under the Surface Mining Control and Reclamation Act of 1977 is a first step in revitalization of university involvement with U. S. Mineral Technology.

It comes as no surprise to this audience that universities are not independent institutions--although many independent people still make their professional homes in them. The state and federal bureaucracies have intruded extensively into academic programs--both research and teaching programs-- both positively and negatively. In the geosciences, without research dollars from federal and state government no significant, continuing research program can be sustained. Industry support from oil and mining companies for geosciences research in universities is and has always been at a much lower level than, for example, industry support for research in nuclear physics and engineering. Exploration geophysics has received a higher lever of industry research support than petroleum and mining geology. However, industry support for geosciences has never attained a level commensurate with either industry or national needs. Industry, the Department

of Energy, and the U. S. Geological Survey should enter into
joint discussions of research needs and responsibilities in
the geosciences.

Universities do not exist in a vacuum. They are very
much a product of the society they serve. Programs in geosciences will not be adequate until U. S. national policy
determines that they should be supported and funds are appro-
priated to support them. This will not occur until the role
of the geosciences in achieving U. S. energy and materials
policy goals is perceived by policy makers. That an industrial society that consumes and alters the earth needs policy
guidance from earth scientists seems so obvious that it
should not need to be said but, regrettably, never have so
few been needed by so many who don't know it.

3

Hydrocarbon Resources and Related Problems

John D. Moody

"Unlike our reserves of coal, iron and copper, which are so large that apprehension of their early exhaustion is not justified, the oil reserves of the country, as the public has frequently been warned, appear adequate to supply the demand for only a limited number of years ..... For some years we have had to import oil, and with the growth in demand, our dependence on foreign oil has become steadily greater, in spite of our own increase in output. It is therefore evident that the people of the United States should be informed as fully as possible as to the reserves now left in this country for without such information, we cannot appraise our probable dependence upon foreign supplies of oil, on the expanding use of which so much of modern civilization depends.

Fortunately estimates of our oil reserves can be made with far greater completeness and accuracy then ever before."

Sound familiar?

These words were published in the Bulletin of the American Association of Petroleum Geologists in 1922, and were co-authored by a blue-ribbon panel of experts convened by the federal government. The panel included such renowned and revered names as W. W. Wrather, Wallace E. Pratt, Alexander W. McCoy, Alexander Duessen, K. C. Heald, W. T. Thom, Jr., Kirtley F. Mather and R. C. Moore.

The committee went on to properly qualify its estimates, and ended by pointing out (and again I quote), "The strong obligations of the citizen, the producer and the Government to give most serious study to the more complete extraction of the oil from the ground, as well as to the evidence of waste, either through direct losses or through misuse of crude oil or its products."

TWO ESTIMATES OF
U.S. CRUDE OIL RESOURCE

	1922	1977
DAILY PRODUCTION *(Million Barrels Per Day)*	1.3	8
CUMULATIVE PRODUCTION *(Billions Of Barrels)*	5.5	112
DISCOVERED RESERVE *(Billions Of Barrels)*	5	50
UNDISCOVERED RESOURCE *(Billions Of Barrels)*	4	150

FIGURE 1

The estimate for the total endowment of crude oil in the United States, made by this committee of outstanding earth scientists in 1922, was 14.5 billion barrels, 5.5 billion barrels of which had already been produced at that time. (Fig.1)

I cite these numbers only to emphasize that "true values" for prospective reserves and undiscovered resources are unknowable, and to express the sincere hope that our present estimate of about 300 billion barrels (of which 112 billion barrels have now been produced) will in time turn out to have been way low.

In any event, the solution proposed for the perceived crude oil crisis in 1922 included eliminating waste, increasing efficiency in usage of oil, end-use control, research on synthetic crude from coal, development of a shale oil industry, more efficient and widespread use of coal, and increased recovery of crude oil from the reservoir. These words also sound familiar, and well they might as they are among the main points in President Carter's proposed energy policy now being debated in Congress.

Not only was the proposed solution to the perceived energy crisis in 1922 quite similar to the current Administration's proposal to alleviate the current crisis, the two proposals are alike in another important respect:

Neither proposal advocated increased exploration and drilling as a means of ameliorating a predicted shortage of hydrocarbons.

But what in fact did happen between 1922 and 1977? How did we get from a producing rate of 1.3 million B/D in 1922 to our current rate of almost 8 million barrels per day?

No shale oil or syncrude industry was created, although intensive research in these important areas has been proceeding vigorously from 1922 to today. Efforts to reduce waste, and lessen demand, and control usage of crude, were largely ineffective. Coal production increased somewhat, and then declined. The only constructive thing that happened was precisely that which was omitted from the proposed solutions to 1922's, and today's, predicted shortages.

Private industry, operating with favorable economic incentives, a free-market price system, and a minimum of government regulations, continued to explore - continued to drill - continued to find new oil fields - built up a natural gas industry - developed a petrochemical industry - fueled a rapidly-growing automobile industry - fueled a near-indispersable aviation industry - fueled the mightiest war machine the world had ever known, which was necessary to end the aggression of Hitler and imperial Japan - provided chemical fertilizers and tractor fuel to grow crops which feed not only our own two hundred plus million people, but millions of hungry people in other countries.

Well, that's a "case history".

Let no one think that by comparing today's situation to that of 1922, I am in any way minimizing the seriousness and gravity of the energy dilemma with which we are presently faced - with which we must cope somehow. But with most knowledgeable estimators agreeing that our undiscovered resource of crude oil in this country is in the range of 100-150 billion barrels, and that our undiscovered resource of natural gas in this country is in the range of 400-600 trillion cubic feet, surely it is worthwhile for the government, and the public at large, to encourage - not discourage - increased exploration, and more drilling. It was the answer in 1922 - it surely is a part of the answer now - and it just may be that, as was the case in 1922, future discoveries might substantially exceed our expectations.

But the energy dilemma is real - it is not contrived - it is not going away.

In the United States, where the petroleum geology is well understood, recent estimates of undiscovered resources of crude oil have been converging on a value of about 100-150 x 10^9 barrels, as I mentioned previously. As techniques used in arriving at these U. S. estimates are applied around the world, no doubt estimates of the world's undiscovered oil and gas resources will become more reliable, but it is clear that reasonable estimates of future reserves and production cannot long fit reasonable projections of demand.

Fig. 2 shows a recent projection of U. S. oil and gas production, and demand, assuming continuation of the current energy policy. (See page 32.)

This next graph shows the effect on oil and gas supply and demand if the major elements of President Carter's energy plan are implemented forcefully. Demand for oil and gas is reduced by increased usage of coal and nuclear power, and by conservation measures; the import bill is still uncomfortably high. (See Figure 3, page 33.)

So where can we turn for alternatives to conventional

oil and gas?

Synthetic fuels derived from fossil fuel resources other than oil and gas such as oil shale, tar sands, and coal are technologically feasible now, but economic and environmental factors involved render useful levels of syncrude and syngas development unlikely, and probably unacceptable, at least within the next ten to twenty years.

Nuclear power is attractive from many points of view, but involves still more environmental insults which many knowledgeable people regard as unacceptable. Breeder reactor technology in any event is not sufficiently advanced to be of any help in the near term, even if the environmental problems can be tolerated. And uranium reserves and undiscovered resources needed to fuel "standard" nuclear power plants appear to be limited, at least in the United States.

Such energy sources as heavy hydrogen, solar energy, geothermal hot rocks and atmospheric electricity, are our "ultimate energy sources," and one or more of these "ultimate energy sources" must be made available on a large scale before we reach the end of our low-entropy, readily-available sources of energy. (Entropy is a measure of the unavailability of energy.)

Research on other energy sources, the most promising of which are fusion and solar, needs to be and is being pursued vigorously, but it is very unlikely that the necessary technology can be made available within the next 25-30 years. This time lag of 25-30 years, needed to develop alternate sources of energy, constitutes what I call the "energy gap".

So we are left with coal.

Coal is by far the most abundant of the fossil fuels. Environmental problems related to mining and using coal are well known, and can and must be mitigated. But given reduction of these problems to acceptable dimensions, the world's coal resources are sufficient to support the world economy at a modest growth rate for several generations.

The world's discovered crude oil resource (proved plus prospective reserves plus production to date) is estimated at about $1,105 \times 10^9$ barrels. Of this amount, 320×10^9 barrels have already been produced, and some 785×10^9 barrels constitute the world's remaining reserve of crude oil, of which well over half is in the Middle East.

The world's undiscovered but potentially

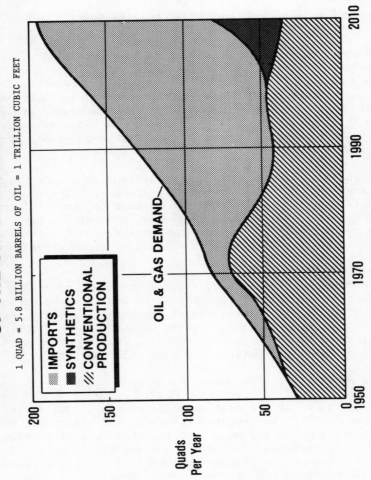

REFERENCE PROJECTION OF THE OIL & GAS SECTOR

1 QUAD = 5.8 BILLION BARRELS OF OIL = 1 TRILLION CUBIC FEET

IMPORTS
SYNTHETICS
CONVENTIONAL PRODUCTION

OIL & GAS DEMAND

Quads Per Year

200

150

100

50

0

1950   1970   1990   2010

Source: Roger F. Naill "Managing The Energy Transition"

FIGURE 2

COMBINED POLICIES PROJECTION
FOR THE OIL & GAS SECTOR

1 QUAD = 5.8 BILLION BARRELS OF OIL = 1 TRILLION CUBIC FEET

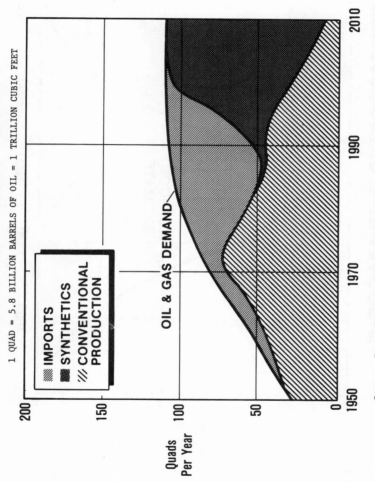

Source: Roger F. Naill "Managing The Energy Transition"
FIGURE 3

WORLD CRUDE OIL RESOURCES DISCOVERED & UNDISCOVERED

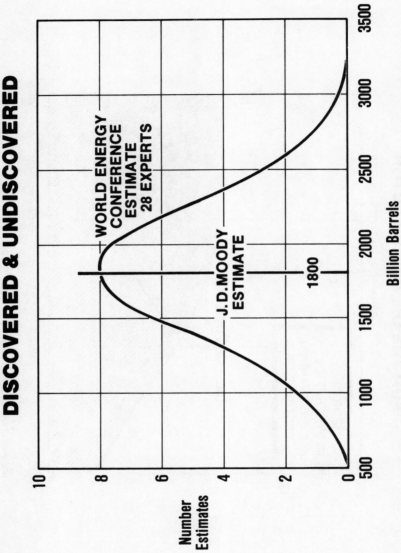

FIGURE 4

discoverable and recoverable resources of crude were estimated by me at about 925×10^9 barrels. This figure is the expected value of a probability distribution whose 5% value is $1,400 \times 10^9$ and whose 95% value is 600×10^9 . (Fig. 4)

This graph presents the results of a recent survey by the French Institute of Petroleum for the World Energy Conference. Some 28 different estimates went into this study, and I am pleased to point out that my own current effort is near the median value of all these experts.

Figure 5 is a projection of cumulative discoveries and production plotted against time, with the asymptote representing the collective opinion of the IFP's 21 estimators for conventional oil resources.

Fig. 6 shows a projection of annual production for the world under three demand assumptions; world production is expected to peak in the decade 1985-1995.

Similarly, the world's undiscovered but potentially discoverable and recoverable resources of natural gas are estimated at about 5200 trillion cubic feet.

Such estimates of essentially unknowable quantities are of necessity subjective, but are useful if not essential exercises in guiding such important functions as exploration, research, and public policy. But whatever the magnitude of the world's undiscovered hydrocarbon resources may be, only a portion of those resources will in fact be discovered and produced. The challenge is to maximize the percentage of undiscovered resources that will be made available to fill human energy needs, and this is clearly a function of economic incentives for both exploration and production amongst other things.

There is no question but that the world is going to need all of its hydrocarbon resources, all of its coal resources, and all of its uranium resources. These are valuable assets and must be recovered. There is also no question but that research aimed at alternative sources of energy needs to be strongly supported, because in time even our coal will be exhausted.

Fig. 7 attempts to illustrate percentagewise the distribution of the low-entropy energy sources which we are now using, around the world. I think it's extremely significant to note that limited contributions in terms of total BTUs available to us in the world of oil and natural gas liquids, gas, tar, and oil shale. They total 10 percent of the world's

CUMULATIVE WORLD DISCOVERY & PRODUCTION

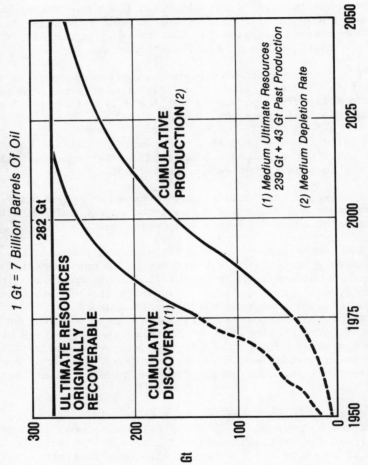

1 Gt = 7 Billion Barrels Of Oil

300

282 Gt

ULTIMATE RESOURCES
ORIGINALLY
RECOVERABLE

200

CUMULATIVE
DISCOVERY (1)

CUMULATIVE
PRODUCTION (2)

Gt

100

(1) Medium Ultimate Resources
239 Gt + 43 Gt Past Production

(2) Medium Depletion Rate

0

1950   1975   2000   2025   2050

Source: World Energy Conference

FIGURE 5

WORLD MAXIMUM PRODUCTION CAPACITY
Medium Ultimate Recoverable Resources: 239 Gt

1 Gt = 7 Billion Barrels Of Oil

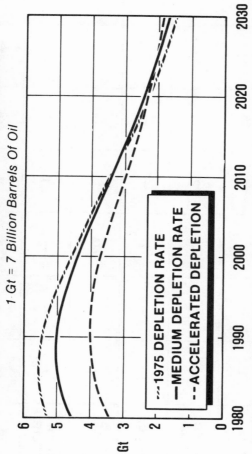

- --- 1975 DEPLETION RATE
- — MEDIUM DEPLETION RATE
- -- ACCELERATED DEPLETION

Source: World Energy Conference

FIGURE 6

DISTRIBUTION & RELATIVE AMOUNTS IN PERCENTAGES OF ENERGY RESOURCES

	Oil & NGL	Gas	Tar	Oil shale	Coal	Uranium	Total
RUSSIA, CHINA, ET AL.	1	—	1	—	26	NA	29
UNITED STATES	—	—	—	2	18	15	35
CANADA	—	—	1	—	—	9	10
MIDDLE EAST	1	—	NA	—	—	—	1
OTHER FOREIGN	1	—	—	1	4	18	24
TOTAL WORLD	4	1	2	3	49	41	100

— None Less than 1%. Some Columns Do Not Add Due To Rounding Total Endowment From Those Sources $\cong 8 \times 10^{20}$ Btu

FIGURE 7

endowment of BTUs. And the rest of the world's available low-entropy BTUs are divided between coal and uranium. Roughly 30 percent of the world's readily available low-entropy energy is in Russia and China, 45 percent is in North America, in the United States and Canada, and in all the rest of the world there is about 25 percent. The Middle East has roughly one percent of the world's energy - yet this one percent is concentrated in the world's largest oil fields, in very low-entropy energy sources. Consequently, the Middle East is able to quadruple the world's total low-entropy energy. Notice the contribution of natural gas -- for all the world the natural gas percentage is about one percent.

These numbers are certainly thought-provoking, when we remember that powerful economic pressures are always forcing us to use our lowest-entropy energy first. The rate at which we are using our oil, our gas, and our natural gas liquids has to concern us, and we are faced with shortages in these commodities. But we are not really faced with an overall energy shortage; we are faced with an energy availability shortage because of the economic pressures that force us to use our low-entropy energy first. And this basically is the "energy dilemma" with which we are faced -- with which we must cope somehow or another.

The "energy dilemma" is beginning to be understood by numbers of people, but unfortunately still not enough <u>really</u> understand. What is even more dismaying is that only a very few people understand that this "energy dilemma" is just a part of a much greater problem. That problem is numbers of people. People produce more people. People require more and more natural resources. People possess rising expectations as far as quality of life is concerned, and people pollute.

The "people problem" is the crux of what I call the "ecology dilemma".

In 1850 the world population was about one billion people. In 75 years this population doubled, so that in 1925 we had two billion people. The next doubling period was 50 years, from 1925 to 1975, and the world population grew from two billion to four billion people. It is estimated that the next doubling period will only be 35 years, so that by the year 2010 there will be eight billion people in this world. That is, in about 35 years - the duration of our energy gap - for every person in the world now, there will be two people. And this rate of population growth seems to be pretty intractable - there's not much that can be done about stemming this increase. So we have to consider our energy problems

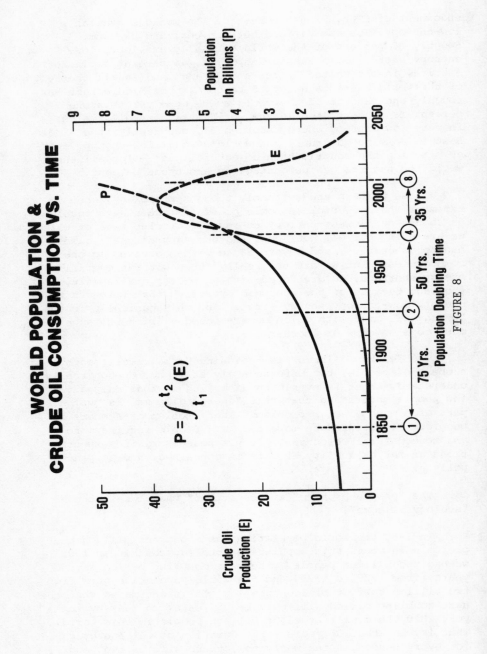

WORLD POPULATION &
CRUDE OIL CONSUMPTION VS. TIME

$$P = \int_{t_1}^{t_2} (E)$$

FIGURE 8

and our environmental problems in the light of this growth
in population.   (Figure 8).

The period 1850 to 2010 coincides roughly with the advent and growth of the oil and gas business, and during this
period of unprecedented population growth from one billion
to four billion people at the present time, and a projected
eight billion people in 2010, we're looking at the period in
human history when oil and natural gas have become major
factors in our economy. We also have witnessed the development of a high-technology society in the same time span.
What is the interrelation between this rapid population
growth, the advent of a new source of energy -- oil and gas
-- and the tremendous technological advances in our society
which have improved life quality substantially? Which is
cause and which is effect? I submit to you that it's the
improved availability of energy that has enabled our high-
technology society to sustain the tremendous population
growth we have experienced, and at the same time enabled us
to enjoy a substantially increased life quality. Prior
to 1850, prior to the advent of our modern oil and gas industry, the world's population rocked along at somewhere
around a billion for several centuries, anyway. So the
development of the technology on which we thrive, and on
which we are becoming so reliant, and which we are so
accustomed to take for granted, was really made possible
because of the availability of cheap energy.

But cheap energy is a thing of the past. And in the
year 2010, when the world's population will be pushing
eight billion, our oil and natural gas will most likely be
used up, so we can only hope that the energy gap will have
been circumvented in some manner.

Human population, quality of life, environmental degradation and natural resource utilization are intimately
interdependent and interrelated. As population increases,
so does environmental degradation. Increased natural resource utilization is essential to maintaining life quality
for an increasing population. If natural resource utilization is decreased even with a steady population, quality of
life will inevitably decline. These inescapable trade-offs
can be simply represented as shown in Figure 9, where L
is human life quality, P is human population, NR is natural
resource utilization, and ED is environmental degradation.
(See Figure 9).

In the past we have relied heavily on technological im-
provements-or what I call quick technological fixes, or "QTFs"
in natural resource utilization to solve our problems for us-

FIGURE 9

$$L = f \frac{\overline{NR}}{P \cdot \overline{ED}}$$

L = **LIFE QUALITY**

P = **POPULATION**

$\overline{NR}$ = **NATURAL RESOURCE (Energy) UTILIZATION**

$\overline{ED}$ = **ENVIRONMENTAL DEGRADATION**

to improve our life quality. We've been quite successful at
this. However, we are running into physical limits of resource availability, physical limits in terms of population
growth and degradation of the environment, and consequently a
reduction of the capability of the land to support life. The
world's human population is in direct competition with all
other populations.

Natural resource utilization is largely energy utilization, and consequently the energy factor implicit in the
equation is of the greatest interest to us here. We can see
that as we approach physical limits of natural resources,
and particularly energy, all the other problems are exacerbated. This, then, is a simple statement of the "ecology dilemma." There is simply no getting away from the interrelation of these four factors of population growth, energy utilization, environmental degradation, and life quality.

An historical example of the "ecology dilemma" was
cited in the New York Times of November 3, 1975, and this is
my second "case history."

"Ye Olde Pollution Problem"

"Rising fuel costs and urban air pollution are not, as
logic might seem to dictate, the undesirable by-products of
the industrial revolution, grown to giants in the 20th century. According to evidence adduced recently by William H.T.
Blake, a historian, those city ills were already evident in
Medieval England."

"The cause of the environmental crisis in the mid-13th
century was the depletion of the country's woodlands; the
reason, the growth of England's population from 1.1 million
in the Doomsday survey of 1086 to an estimated 3.7 million by
the early 14th century."

"To support the growing population, woodlands were converted to arable fields, and the familiar modern price snow-
ball began. By 1255, according to the Calendar of Patent Rolls
of Henry III, wood consumed in forges was of more value than
the iron the forges produced; and coal (from Newcastle, and
elsewhere) became the fuel of choice in London's hearths and
industries. By 1288, a royal commission found 'the air...
infected and corrupted' in industrial areas. In 1307, pollution controls and penalties, were enacted."

"What righted the imbalance between population growth
and resources were famine and the Black Death of the 14th
century. Not until the mid-17th century was air pollution

U.S. ENERGY CONSUMPTION

1 QUAD = 5.8 BILLION BARRELS OF OIL = 1 TRILLION CUBIC FEET

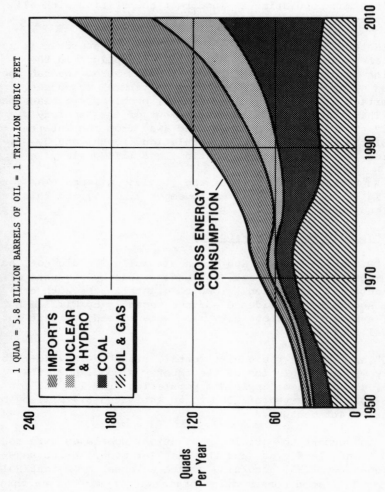

Source: Roger F. Naill "Managing The Energy Transition"
FIGURE 10

again regarded a serious nuisance, and 'Presumtuous Smoke'
considered a health hazard."

This is what I call the "Black Death" solution. It was
brought on at least partly by a "laissez-faire" attitude, by
inadequate response--too little too late--from government.
Hopefully this can be avoided in today's context.

What restraints can we forsee -- are we now in fact ex-
periencing -- with respect to our efforts to mitigate the
"energy dilemma"? I have already discussed the powerful restraint imposed on us by a relentless exponential growth in
population. And although attempts to quantify the physical
limits of our low-entropy energy sources are mere educated
guesses, we know beyond the shadow of a doubt that there are
such limits, and in the United States there is substantial
evidence that we are approaching these limits for oil, gas
and U_3O_8. Other more immediate restraints are politico-
economic, and are more frustrating because they are imposed
on us by governmental fiat at precisely the point in time
when our efforts to find that next barrel of oil, that next
cubic foot of gas, that next ton of coal and that next pound
of U_3O_8 should be encouraged by society -- not discouraged.

I'd like to quote from a recent discussion of these
problems -- "Managing the Energy Transition," by Roger Naill.

"During the next few decades, the United States must
undergo a transition from dependence on domestic oil and gas
to reliance on new sources of energy. (Fig. 10)

"Twice previously in its history the United States has
restructured its economy to move from primary reliance on one
fuel to another; the transition from wood to coal, and from
coal to oil and gas, have each taken over 60 years to complete. Because each transition introduced a more efficient,
convenient form of energy, past transitions have encouraged
growth in the other sectors of the United States economy.
But the current transition is different -- there is no cheap,
abundant, and environmentally acceptable alternative to the
use of domestic oil and gas. For lack of a domestic alternative, the United States is currently following a path that
has led to excessively high imports during the early transition progresses. To counter this trend toward imports, the
federal government is currently considering the following
proposals:
- mandatory conservation measures
- oil import quotas or tariffs
- decontrol of oil and gas prices
- accelerated development of nuclear power

- lower air quality standards
- accelerated synthetic fuels research and developement
- federal loan guarantees or price supports for synthetic fuels commercialization
- rate reform for electric utilities
- federal subsidies for coal development

In each case, government policymakers must determine whether federal intervention is warranted: benefits from lower imports must be weighed against the economic and environmental costs of each program. To analyze the long-term effects of federal policy on the United States energy system, a system dynamics model, entitled COAL2, has been constructed at Dartmouth College's Thayer School of Engineering. This model emphasizes the complex causal mechanisms that determine the production and consumption of energy over the long term. The model has been used to assess the relative effectiveness of the alternative energy strategies currently under consideration by United States policymakers.

"The model study concludes that the United States energy problem is fundamentally more severe than other analyses and agencies indicate, and much less amenable to a near-term solution. More specifically,

- Given no major changes in energy policy, the United States must import almost 40% of its energy from 1985 to 2000. (Fig. 10)
- If energy demand stabilized by the year 2000, imports would peak in 1985 at 33 quads per year (34% of consumption), and drop to zero by 20.0.
- An Accelerated Nuclear program is ineffective in reducing United States dependence on imports; with it, are reduced only 10% below the base case projection by the year 2000.
- An Accelerated Coal program could achieve the national goal of independence from foreign oil imports, but not until the year 2000. Yet this program could cause severe economic dislocations due to the major shifts in investment to the energy sector.
- A combined program that stabilizes energy demand over the long term and accelerates the use of coal (the Zero Energy Growth and Accelerated Coal programs) generates a smooth transition that balances United States energy supply and demand by the year 2000.
- Under any policy circumstances, United States dependence on imports increases to 1985, energy prices increase substantially above 1975 levels, and much investment -- and foresight -- will be required of citizens, government, and industry during the next 35 years.

"United States energy policymakers are the first to admit that there is no easy solution to the U. S. energy problem. Our analysis suggests that <u>the difficulties in establishing an effective U. S. energy policy are an inherent property of the energy system</u>, not simply a case of politicians unable to make up their minds. It is a general property of complex systems that there is often a fundamental conflict between the short-term and long-term consequences of any policy change. In the U. S. energy system, <u>any</u> energy policy action creates a controversy between short-term and long-term interests.

"For example, the COAL2 model indicates that the policy changes that improve the long-term energy picture are likely to have negative short-term side effects. Energy prices, capital investment, and environmental damage all increase substantially before 1990 if Zero Energy Growth and Accelerated Coal policies replace the trend projections of the reference run. Such is the price the system exacts for the long-term benefits of reduced imports that can accrue only after 1990. (Fig. 10)

"Conversely, policies designed to improve the behavior of the energy system over the sort run will often worsen our energy problems over the long run. The reference run of the COAL2 model, representing current U. S. energy policy, reflects this property. In an effort to continue to supply U. S. consumers with low-cost energy in the short run, current energy policy will almost certainly lead to a post-1985 period of uncertain supplies, excessively high energy prices, and political and economic disruption due to massive dependence on imports.

"It is an unfortunate fact of the current political decision-making process that short-term considerations most often dominate policy considerations. For example, the oil price rollback called for in the Energy Policy and Conservation Act of 1975 was designed to avoid the short-term shock to the U. S. economy of a sudden increase in energy prices due to decontrol of old oil. The short run is always more visible and compelling. It speaks loudly for immediate action. Yet a continuation of current policies aimed at short-term improvement could lead to long-run economic and political consequences -- severe recession, energy shortages, or a war for the control of Arab oil -- that are totally unmanageable and unacceptable.

"We conclude that the nation's long-term interests are best served by reducing energy demand and accelerating the use of coal, in spite of the short-run economic and environ-

mental sacrifices that accompany these policies. Even if
these policies are not carried out exactly as suggested, we
hope that our modeling effort has clarified the long-term
U. S. energy future and disspelled some common misunderstand-
ings about the U. S. energy problem. If the model can
achieve this, long-term considerations will weigh more heavily in the decision-making process -- as they properly should."

As Bill Fisher, recently of the Department of Interior,
likes to point out, we "are going up the down escalator" on
energy matters in this country and most other countries.
And, in the absence of a series of unprecedented and unantic-
ipated QTF's which would enable large-scale utilization of
one or more of the ultimate energy sources, we in the United
States have no viable alternative to increased dependence on
foreign oil and gas for the duration of the energy gap, mitigated only by such amounts of hydrocarbons as we are fortu-
nate enough to find, or are permitted to find, within our
borders.

The capital and economic consequences of this situation
are enormous, to say nothing of the implication with respect
to the "ecology dilemma". Something akin to the "Black
Death" solution can and must be avoided. It can be avoided
if the positive aspects of President's Carter's proposed
energy policy, augmented by the "de facto" solution of 1922's
perceived crises -- namely increased exploration and stepped-
up drilling, are implemented promptly, vigorously, and forcefully.

4

Coal

Jack A. Simon

Introduction

Coal, after many decades as the principal source of
energy in the United States, for the first time contributed
less than 50 percent of total energy in 1946. The percentage
has generally declined, reaching a low of 17.2 percent in
1975, to the present level of about 19 percent of the total
energy used in the United States. There is a high degree of
confidence that coal can meet its current share of national
energy needs and also can increase its share to take up some
shortfall in other energy sources. There is a national com-
mitment to double coal production in the decade 1976-1985.
This is certainly more than normal growth for such a period
and is intended to offset present and near-future shortfalls
in domestic oil and gas supplies, and delay in construction
of nuclear power plants.

Although the geosciences have played a significant role
in past delineation of coal resources, geoscientists have
been employed less in the coal industry than in any other
energy-mineral industry. The accessibility of coal, which
has been widely found at and near the surface in many parts
of the country, has permitted much development without major
help from the geosciences. From personal experience, however, I know that geosciences could have been used more extensively and beneficially. Discovery and delineation of
other mineral resources have generally been more difficult
and thus have been more dependent on applications of science.

I believe that now, and in the future, the geosciences
have substantially more important contributions to make in
utilizing coal as an energy source. These contributions are
related to evaluation of resources, mining geology, utilization of coal, and environmental consequences of development
and utilization of coal.

Preliminary data indicate that U. S. coal production constituted nearly 19 percent of the world's production of coal in 1977. The United States is estimated to have 20 to 30 percent of the coal resources of the world and a similar percentage of reserves.

Coal Resources and Reserves

A major impact of application of geosciences and particularly geology in the past has been delineation of areas of coal occurrence. For more than 75 years many estimates have expressed quantities of coal as resources and reserves within variously defined limits.

It has become essential that better defined and realistic presentation of coal resources and reserves data be made so that decision makers, who are rarely earth scientists, and the public can better understand the significance of quantities of coal reported in geologic studies.

By definition, coal resources represents coal in the ground that might ultimately be recovered. In practice, however, estimates of coal resources have generally included coal in the ground that exceeds a defined, relatively thin minimum thickness which is sometimes further classified in broad categories of depth. The first prerequisite for developing meaningful numbers relative to coal which might be developed to meet our needs, of course, is that the coal is present. Data on coal resources, however, generally provide information of limited significance for near-term energy planning. Data on coal reserves are of greater importance for such planning.

In practice, estimates of coal reserves, which also have been defined in several ways in different studies, have attempted to apply to coal resources more stringent restrictions related to practical mineability. Most commonly this more stringent restriction has specified a greater minimum thickness than was specified for resources, and sometimes the definition included more restricted depth ranges.

To have, for the whole country, estimates of coal in the ground that are technically, economically and legally minable by present mining practice would be highly desirable, if practical. Although this is increasingly an objective of coal reserves studies, many types of essential data for such an assessment are not available for broad areas.

There remains a major contribution to be made by geoscientists to provide more refined coal reserves data for

effective energy planning. It is my strong personal feeling
that persons reporting or using coal resource or reserves
data do a great disservice when they divide a nebulous resource or reserves number by current annual production to
yield a number that purports to suggest the number of years
that coal will last. Estimates of even "recoverable reserves"
are still too inexact for such applications to have meaning
now or at some future date when, in fact, coal needs may be
a significant multiple of current production. There remains
a great need for more refinement of data on recoverable coal
in each coal area in the country.

Among factors which must be considered are thickness of
seam, depth, coal quality, geologic character of overlying
and underlying strata, associated geologic features which
may influence mining, availability of land for mine development, markets and potential markets in terms not only of
coal users but location in the country relative to the area
where coal needs exist, environmental impacts of mining, and
utilization and disposal of wastes. Each of these are too
complex to detail here, but are suggestive of the kinds of
required data on recoverable coal that are needed for sound
energy planning. Geosciences can provide the basic data or
support data for such assessments.

Before leaving a discussion of resources and reserves,
I would like to leave you with an overall perspective.
Figure 1 shows the major known coal-bearing areas of the conterminous United States. In general, the large reserves of
higher rank bituminous coals (including most coking coals)
are found in the eastern and central U. S. coal fields. Although coals of these ranks are found in western states, the
very large resources there are lower rank lignite and subbituminous coal. The coal fields have been somewhat arbit-
rarily divided into Eastern U.S., Central U.S., and Western
U.S.

In Figure 2, an attempt has been made to generalize some
aspects of recoverable coal reserves based on best data now
available. For purposes of this discussion, there is considerable generalization of the data shown. In Figure 2, the
full circle represents recovered and best current estimates
of recoverable coal reserves for the conterminous U.S. as of
January 1974 adjusted to 1975 by the use of later data. Of
the estimated original recoverable reserves of about 263
billion tons (exclusive of Alaska) remaining recoverable
reserves, shown in solid pattern, total about 225 billion
tons. The hatchured pattern for each major coal field area
of the country shows the portion of original recoverable
reserves which has been mined, a total of 37.15 billion tons.

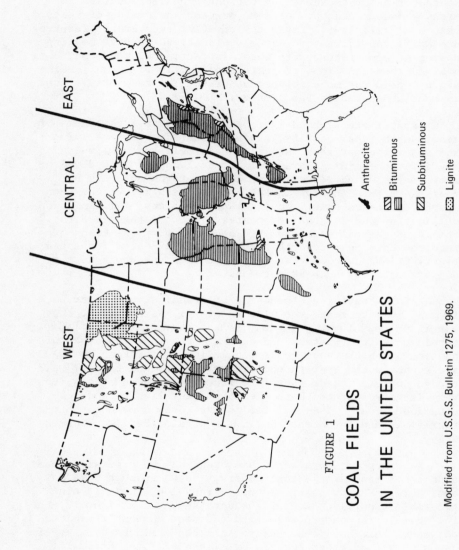

EAST

CENTRAL

WEST

Anthracite
Bituminous
Subbituminous
Lignite

FIGURE 1

COAL FIELDS
IN THE UNITED STATES

Modified from U.S.G.S. Bulletin 1275, 1969.

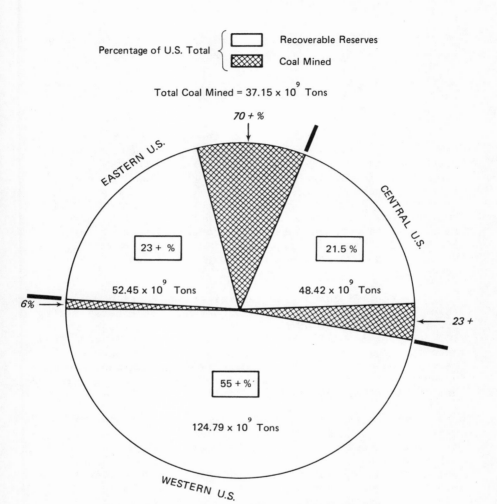

RECOVERABLE COAL RESERVES AND COAL MINED

1975

(Exclusive of Alaska)

FIGURE 2

Percentages shown are percentages of U.S. totals - remaining reserves in the solid pattern, and percentage of total U.S. production in the hatchured pattern.

The purpose of presenting these data is to demonstrate at least a few of the problems in analyses of the most definitive information available on quantity of coal remaining. As noted previously, however, even in this presentation, it is not practical to present detailed information in each part of the country.

As noted on Figures 1 and 2, the eastern states generally have the higher rank coals. These have been near major industrial areas and areas of high population density. The importance of these relationships is reflected by the fact that 70 percent of total U.S. production has come from this area. Note, however, that approximately one-third of the original recoverable reserves in this area have been mined. As development generally occurs in the most favorable deposits first, a substantial part of the remaining coal will be less favorable for recovery than the coal that has been mined.

The coal fields of the central states contain recoverable reserves, mostly of bituminous rank, but generally of lower rank than most of the eastern coals. Although remaining reserves are only slightly less than in the Eastern area, note that only about 15 percent of original recoverable reserves have been mined. Through 1976 the bituminous coal production in the eastern half of the country has constituted nearly 94 percent of the U.S. total.

Although coals of all ranks are found in fields of the western states, about 88 percent of reported recoverable reserves are found in generally thick and shallow deposits of subbituminous coal and lignite in Montana, Wyoming, and North Dakota. These coals are of generally lower quality by most standards. Because they have relatively low sulfur content, however, development has been expanding rapidly to serve increasing needs locally and needs particularly in the central states where many facilities have turned to western coals for use in their plants in order to meet EPA sulfur emission regulations.

Over 55 percent of currently estimated recoverable coal reserves lie in the western states. Although they have contributed only about 6 percent of past total U.S. production, western states in 1976 contributed nearly 19 percent of national production and that percentage is projected to increase significantly by 1985. Most of remaining recoverable

reserves in the eastern and central states will be recovered
by underground mining, although significant surface mining
production will continue for a number of years. In western
states, the largest percentage of coal reserves now considered recoverable will be mined by surface mining.

Coal must be mined, processed, shipped to a point of use
and be converted to energy in order to help meet energy
needs. For expansion, commonly 5 to 10 or more years are required to complete the process. Actual or estimated presence
of coal in the ground does not automatically guarantee that
our needs will be met. Data on recoverable coal reserves in
any area are simply a reflection of potential in that area
and do not indicate the quantity of coal that will be producible at any given time.

Geology of Coal

For many years, geoscientists have studied broad aspects
of the geology and geochemistry of coal and associated strata
including the origin and character of coal seams, environments of accumulation of coals and associated strata, coal
metamorphism, physical and chemical properties of coal,
mineral matter in coal, properties related to specific coal
uses, and certain properties of coal ash.

Until about 15 years ago, a relatively small number of
geoscientists were engaged in coal work, with the possible
exception of delineation of coal resources and reserves.
Even within the coal industry, use of geologists for exploration was limited with most such work being carried out prin-
cipally by state geological surveys, the U.S. Geological
Survey, and the U.S. Bureau of Mines in some areas.

Need for geologic data beyond exploration and estimates
of reserves is widely recognized at present in the coal industry, in many governmental agencies and in a number of
universities. For a number of decades, geologists have been
called on only sporadically for solution of specific problems that have been encountered in mining, processing, or
utilization of coal. The large amount of data developed
through the years by a variety of geoscience investigations
provides a vital data base for more direct application in
solving problems related to major increase in coal production and utilization.

The geologist has unique capabilities for making contributions in some key areas. Of foremost importance is
utilization of geology in underground mining. As we have
seen from an assessment of past production and present re-

serves, the major share of coals in the eastern half of the
country will be mined by underground operations.

Geology must have much wider application in underground
mining in the future. There are a number of reasons that
account for recognition of greater needs for use of geology.
Installation and operation of underground coal mines are
much more costly than formerly, and geologic conditions
that may adversely affect mining operations must be more
effectively handled.

During a period of about 25 years following World War
II, measurement of productivity of individual miners showed
rapid increases. For the past 8 years, productivity has declined dramatically. There are complex reasons for this decline but a major factor has been related significantly to
new laws and regulations affecting mining. Since coal will
be expected to provide a significantly increasing share of
the energy burden, it is important to achieve increased
productivity, at least approaching the 1969 level, in each
mining area.

Certainly many of the health and safety regulations
which have reduced productivity are not likely to be significantly relaxed. No radically new mining technologies that
may ultimately evolve appear to be imminent. Thus, a major
area of attack on the problem of mining productivity can and
should be a greater knowledge of the natural conditions in
which a mine operates in order to (1) better understand and
more effectively deal with geologic conditions which have adversely affected mining and (2) anticipate potential problems related to variable geologic conditions.

The application of this broader use of coal mining
geology supported by geochemistry and geophysics is in a relative stage of infancy in the coal industry. Despite the
fact that you may know a geoscientist who has been making
such applications of geology for many years, such use applied
to coal mining is relatively new in the United States.

Other applications of geoscience related to coal mining
have been more widely applied for a number of years, but
again the actual number of geoscientists involved, until
recently, has been relatively small. Some applications of
geoscience, related both to mining and utilization of coal,
include investigations of variation in coal quality as measured by heating value, ash content, sulfur content, character and distribution of mineral matter, and other variations
in chemical and physical properties of coal as observed within a single mine or within a mining district.

Great potential exists for much broader applications of geophysics, which has been so successfully used in oil exploration and in a more limited way, for coal exploration. Broader applications of geophysics in the mine for detecting weak mine roof, locating faults or other major disturbances in the seam mining hydrology, and sources of coal-associated gas are needed. Greater incentives may now exist since such applications should favorably improve coal mine productivity.

Geochemistry and Physical Properties of Coal

Coal chemistry has a very long and distinguished history of accomplishment. Applications of chemistry to coal with recognition of geologic controls have received increasing interest in recent years. Although the line between coal chemistry and coal geochemistry may not be well defined, recognition of geologic factors which are responsible for the complex characteristics of coal has been an important feature of geochemical research.

A handful of geoscientists through the past 60 years or more have directed attention to some physical properties of coal as seen under the petrographic microscope. The very wide range of chemical and physical properties of coal constituents recognized by coal petrologists have favorably influenced geochemical studies. Applications of coal petrography have been successfully applied to prediction of coking character of coal and of coal blends. The application of coal petrology to specialized uses of coal such as coal gasification, liquefaction, or direct use as a source of chemicals has become an increasingly important area of research.

Of particular recent awareness and concern are the mineral constituents associated with coal. Application of relatively new techniques permits determination of the mineral constituents in coal as opposed to high-temperature ash resulting after combustion. Some noncoal elements in the mineral matter in coal may be volatilized and be emitted with stack gases, and others may be found in fly ash or other ash or process wastes. The form and distribution of minerals in coal are being studied to provide a basis for evaluation of the pollution potential of gaseous, liquid, and solid wastes from coal processing. The characteristics of mineral constituents may affect use, as in combustion or other processing, and some minor and trace elements may be catalysts or may poison catalysts in chemical processing of coal. The mineral content of wastes from coal mining and processing may contain a sufficient amount of minerals that are in short supply to constitute a new source for such minerals.

Although there has been a great expansion in research on mineral matter in all major coal producing areas of the United States in the past five years, most of the work to date consists only of data collection. Comprehensive synthesis of such data should yield greater understanding of the characteristics and distribution of mineral constituents in coal.

Coal and the Environment

In an assessment of the impact of the geosciences on critical energy resources, the impact of geology on certain environmental problems may be equally as substantial as it will be in estimating recoverable reserves, coal mining or utilization of coal.

Underground mine operations must deal with problems of surface subsidence, proper disposal of mine wastes and refuse, and potential problems of disturbance or pollution of ground or surface waters. Surface mine operations face many of the same problems plus the problem of more extensive disturbing of land surface and soil.

Geosciences can provide information necessary for disposal of solid wastes from mining or utilization of coal in an environmentally acceptable manner by determining the characteristics of the wastes and identification of adequate disposal sites. Geoscientific data are needed in solving problems of disposal of liquid or gaseous wastes. The occurrence and forms of sulfur in coal and methods of reduction of sulfur in coal continue as major objects of geoscience research.

Summary

To briefly summarize this presentation on coal:

1. Coal will be a major contributor to our total energy needs for the foreseeable future.

2. There are large reserves of coal in the United States that can sustain an increase in coal production and much larger resources that may be ultimately minable. Resources and reserves data now available have value if properly used. Much greater refinement of data on currently recoverable reserves in each mining area is essential for needed energy planning.

3. To contribute to meeting energy demands, coal must be produced, moved to sites of use, and utilized.

4.   There are many constraints to coal production.
Problems of mine safety, productivity, geographic location,
and environmental constraints on productivity and utilization of coal are all factors that will significantly influence expanded coal use.

5.   Solution of many current problems is necessary not
only to maintain the current percentage of total energy
production contributed by coal, but to increase production of
coal to substitute for at least a portion of the current and
projected shortfall in available gas and oil.

6.   Coal has a vital role as an energy source to assure
national well-being. The geosciences have a vital role to
play, to a degree far greater than at any previous time, to
assure that coal can fulfill its perceived role in meeting
future energy needs.

5

Geothermal Resources

William L. Fisher

In assessing the impact of the geosciences on critical
energy resources, we can observe in the case of geothermal
resources that at least some impact has been registered in
the name itself. Actually, when we view certain of the current, popular notions about geothermal energy and its potential, one is tempted to say that may be the major geoscience
input. Certainly, some of the exceptionally high potential
assigned to geothermal resources gives credence to such observation.

The geothermal energy of the earth is diffuse, diverse,
and by any measure immense. Geothermal resources are generally classed in three main categories: (1) hydrothermal systems, including both vapor-and-liquid-dominated systems, (2)
hot dry rock systems, and (3) geopressured systems.

Geothermal is also popularly classed as an "alternate"
energy resource. But like many other alternate energy
sources, geothermal energy has been used for a long time.
Its use in space heating and agriculture spans recorded time
and in many places of the world. Its use in Iceland and
Southern Russia for these purposes is well known.

Where geothermal energy exists as steam or very hot water, it is used to generate electricity. Its use in electrical generation dates to 1904 in Larderello, Italy. Since
1960 geothermal has been used in power generation at the
Geysers in California. In both cases the energy is recovered
as dry steam. At other places, such as New Zealand and Cerro
Prieto, Mexico, high-temperature water is used; steam is derived by passing the water through a flash separator.

In many deposits, water temperatures are too low for efficient flash separation. In such instances a different geothermal power process-- the vapor turbine or binary system

has been developed. Water brought to the surface is passed
through a heat exchanger where heat is used to boil and super
heat a high-density vapor. The vapor then expands through a
turbine to produce power, and is then recycled to the heat
exchanger.

The largest volume of geothermal waters are too cool for
power production. Past and future use is largely limited to
local space heating and certain agricultural uses.

The second category of geothermal resources--hot dry
rock systems--obviously involve neither steam nor water.
Techniques are being tested whereby fractures are created in
hot dry rock, water is pumped into the cavities, and recovered at other wells as steam.

A final category of geothermal resources is the abnormally pressured or geopressured zones that exist in certain
sedimentary basins of the world. Perhaps the best known is
the Gulf Basin of the U.S. and Mexico. Such abnormally high
fluid pressures develop when interstitial formation waters
are trapped during burial and subsequent basin subsidence.
If the contained water cannot be squeezed out because of
restricted fluid movement, compaction of sediment grains will
not occur and pore water will begin to carry a part of the
overburden load. When this occurs, formation fluid pressures
greater than hydrostatic develop. In sedimentary basins underlain by thin crust, upward flow of heat from the mantle
causes fluids within the geopressure zone to develop anom-
alously high temperatures. The undercompacted or geopressure
zones act as heat insulators; normally pressured zones above
and below these zones have normal temperature gradients.

At the moment, the use of geothermal energy is not very
significant--in terms of total energy supply only a fraction.
of a percent. By 1985 geothermal may contribute 1 percent of
the U.S. electrical power supply and by the year 2000, it
could contribute up to 4 quads of energy, or 2 to 3 percent
of the projected U.S. consumption by that time.

By contrast, the volume of geothermal energy is large
indeed. The U.S.G.S. calculates some 31 million quads of
geothermal energy in place in the U.S., considering heat
above 15° C and to depths of 10 kilometers.

However, estimates of how much of this stored heat constitutes a resource and how much might be recovered for use
differ by at least a thousand times. There are obviously a
number of factors involved in these widely varying resource

estimates. For one, there is a wide range of assumptions relative to technologic capability and future technologic development. There is, in geothermal as with other energy and mineral sources widely different and commonly most uncritical uses of the terms reserves and resources. But probably the major cause of widely varying estimates is the lack of data about the resource itself, or in certain cases, a failure to utilize fully the data that are available.

And in this context I would like to examine the impact, or lack thereof, of the geosciences and do so in the specific case of the Gulf Basin geopressured-geothermal resources.

For here reported estimates of recoverable energy range from 100 quads to 50,000 quads. These large estimates have spawned what might be described as the "Great Methane Debacle".

Why? Because of the lack of geoscience data? Hardly. For it must be remembered that these widely ranging estimates of recoverable energy come from what is possibly the best known geological basin in the world--a basin where over 300,000 wells have been drilled, where hundreds of thousands of miles of geophysical surveys have been run, where something on the order of 150,00 geoscience man-years of experience have been logged. I am not suggesting that all these data are directly applicable nor that even if all these data and experiences were utilized there would be no unanswered questions about Gulf Coast geothermal sources. But I am contending that an adequate use of geoscience data might narrow the range from the present 500 times.

Geopressured zones of the Gulf Basin Wilcox, Vicksburg, Frio and Miocene, underlie and aggregate area of about 330,000 square kilometers. These zones consist of varying amounts of sandstone and shale, containing geopressured fluids with average temperatures on the order of about 300°F, at depths ranging from 9,000 to 22,000 feet (Dorman & Kehle, 1974; Bebout et al, 1975 a,b). Unlike other geothermal areas the energy potential of these geopressured-geothermal zones is not limited to thermal energy. The high pressures are a potential source of mechanical energy, and perhaps most important, the geothermal waters contain dissolved natural gas, chiefly methane, which is potentially recoverable as well.

In 1975, the U.S. Geological Survey (Papadopulos, et al 1975) published an assessment of the Gulf Basin geopressured-geothermal resources as a part of their overall assessment of geothermal resources throughout the U.S. The U.S.G.S. assessed about 145,000 square kilometers or a little less than one-half the area they consider underlain by geopressured zones. The assessed area was divided into 21 subareas with each area considered as a single, idealized "conceptual" reservoir. Critical parameters for these 21 subareas were determined or assumed based on information from 193 wells for the entire region. Thickness of the "conceptual" reservoirs was assumed to equal the extrapolated thickness of the geopressured interval in each area. Each subarea was assumed to be a single sand aquifer, underlain and overlain by two single confining shale beds; and with both sand and shale beds assumed to be continuous throughout the area. Average pressure, temperature, permeability, porosity, and salinity was assumed for each subarea or conceptual reservoir. From these parameters a "fluid resource base" was calculated, and in-place energy values were reported as follows:

Thermal energy	43,000 quads
Methane	24,000 quads
Mechanical energy	220 quads

Various recovery levels of this in-place resource were given based on assumptions of hydrogeologic factors and well spacing. For example, the highest recovery of 3.3 percent, requiring about 26,000 wells with spacing ranging from 1.9 to 2.9 kilometers, gives a well-head yield of energy as follows:

Thermal energy	1,420 quads
Methane	790 quads

Assuming about 8 percent of the thermal energy and 35 percent of the methane energy could be converted to electricity, about 390 quads of the in-place resource could be utilized.

Although a more detailed assessment carried out subsequent to that of the U.S.G.S. indicates that the Survey's estimates of both in-place and recoverable resources may be optimistic, it is interesting to examine how these original estimates have been extended to some of the very high reports of methane potential.

The first step is a paper in 1976 by William Brown (Hudson Institute Research Memorandum #31, also reported in FORTUNE magazine), wherein he refers to the 24,000 quad methane estimate of the U.S.G.S. and their observation that possibly additional geopressured areas in the Gulf Coast are

1 1/2 to 2 1/2 times those assessed by the Survey. That leads Brown to an extension of the methane resource base to 60,000 to 84,000 quads. He goes on to cite Paul Jones' (Jones, 1976) observation that simplifying assumptions in these calculations are frequently erroneous--up to a factor of 2--mostly on the conservative side. With this reasoning Jones ups the methane resource to 52,000 quads, and further assumes that the geopressured shale formations will yield as much methane-saturated water as the sand aquifer (such an assumption was also made by the Survey in their original cal-culation of 24,000 quads). Thus, the methane resource is doubled again, now standing at 105,000 quads (Jones, 1976). Brown's presentation of 60,000 to 84,000 quads seems conser-vative by comparison. The original U.S.G.S. estimate has been quadrupled.

As to the recovery of this 105,000 quad methane re-source; Brown judges a 4 to 5 percent factor minimal, and goes on to assume a "relatively conservative" recovery factor of 10 to 20 percent. Off of the 105,000 quad resource, the recoverable volume of methane thus stands at 10,000 to 20,000 quads (or TCF), up to 25 times the 790 quads estimated by U.S.G.S.

But the upward revisions are not yet complete. For next we turn to a series of editorials appearing in the WALL STREET JOURNAL in April and May (1977), and titled in succes-sion "1,001 years of Natural Gas", "ERDgate", and "The Energy Crises Explained". Centering the editorials basically on a series of gas supply estimates made by the now-famous Market Oriented Program Planning Study (MOPPS), effort of ERDA, they begin to tally the U.S. natural gas supply. Supply not re-sources. First, they cite 216 TCF of proven reserves, plus 230 TCF the U.S.G.S. lists as inferred reserves, plus 285 TCF from the Devonian Shales, plus 600 TCF from Western tight sands, plus between 200 and 300 TCF of coal seam methane. To this approximate supply of 1,600 TCF is added the supply from the Gulf Basin geopressured methane. Now it stands, not at the 790 TCF estimated by U.S.G.S., not the 10,000 to 20,000 TCF of Brown, but a whopping 20,000 to 50,000 TCF. Now this is supply, not resource. The original Survey estimate is now increased 63-fold. In fact, the 20,000 to 50,000 TCF gas supply is divided by the current annual U.S. consumption of natural gas of 20 TCF, and in words of the WALL STREET JOURNAL "is enough to last between 1,000 and 2,500 years."

What was the impact of geoscience data in these figures?

One more quote from the WALL STREET JOURNAL editorial of May 27, 1977: "Against their own better judgment, geologists

do make predictions; their track record is ludicrous. Of
course, it is conceivable that sometimes the inventory clerks
may be right."

Such viewpoint notwithstanding, realistic assessment
does require the inventory clerk--the geoscientists--working
with available data. Such assessment by the Bureau of
Economic Geology, University of Texas at Austin, under con-
tract to ERDA and now DOE, has been underway for three years
and is now near completion. (Bebout et al 1975a; Bebout et
al 1976; Bebout et al 1978a; Bebout et al 1978b; Fisher 1978,
Loucks, 1978). In this assessment data from about 3,000
wells were utilized, (compared to the 193 of U.S.G.S.), several hundred cores from the geopressured zone, were studied
and analyzed, and seismic profiles were used. Sandstone volumes were mapped in detail in all the zones as were the various parameters critical to resource assessment. These findings show that assumptions of parameters made by the U.S.G.S.
for "conceptual" reservoirs are optimistic. For example, the
volume of reservoir sandstones mapped by the Bureau of
Economic Geology is only about one-half that assumed by the
Survey, total reservoir volume is only about 15 percent, average temperatures are about 10 percent less, average porosity values are about 15 percent lower, average pressures are
about 12 percent less, and average salinites are about 55
percent greater. Finally, and perhaps the most critical, extensive core analyses show permeabilities for all areas ex-
cept the upper Texas Coast Frio to average about 2.6 md or
1/10 the values assumed by the U.S.G.S. The net effect of
these findings is to reduce the volume of recoverable energy
from the geopressured-geothermal zone estimated by the G.S.
by an order of magnitude. Nonetheless a substantial resource
appears to exist in the Upper Texas Coast Frio. A prospect
has been delineated and will be test drilled shortly. Test-
ing of this well over a two-year period will provide answers
to questions not solvable by geologic assessment.

The geothermal resources of the U.S. are capable of making a significant contribution to domestic energy supply.
Prudent policy must, however, be based on realistic expectation, which in turn must come from rigorous evaluation of
geoscientific data.

References

Bebout, D. G., Dorfman, M. H., and Agagu, O. K. (1975a) Geothermal Resources--Frio Formation, South Texas: Bureau of Economic Geology, The University of Texas at Austin, Geologic Circular 75-1, 36 p.

Bebout, D. G., Agagu, O. K. and Dorfman, H. H. (1975b) Geothermal Resources, Frio Formation, Middle Texas Gulf Coast: Bureau of Economic Geology, The University of Texas at Austin, Geologic Circular 75-8, 43 p.

Bebout, D. G. Loucks, R. G., Bosch, S. C., and Dorfman, M. H. (1976) Geothermal Resources, Frio Formation, Upper Texas Gulf Coast: Bureau of Economic Geology, The University of Texas at Austin, Geologic Circular 76-3, 47 p.

Bebout, D. G. Gavenda, U. J. and Gregory, A. R. (1978a) Geothermal Resources, Wilcox Group, Texas Gulf Coast: Draft Report to the U.S. Department of Energy, Bureau of Economic Geology, The University of Texas at Austin, 82 p.

Bebout, D. G., Loucks, R. G., and Gregory, A. R. (1978b) Geopressured geothermal fairway evaluation and test-well site location, Frio Formation, Texas Gulf Coast: Draft Report to the U.S. Department of Energy, Bureau of Economic Geology, The University of Texas at Austin, Geologic Circular 74-4, 33 p.

Brown, W. M. (1976) 100,000 quads of Natural Gas?: Hudson Institute, Research Memorandum #31, 32 p.

Dorfman, M. H. and Kehle, R. O. (1974) Potential Geothermal Resources of Texas: Bureau of Economic Geology, The University of Texas at Austin, Geologic Circular 74-4, 33 p.

Fisher, W. L. (1978) Texas Energy Reserves and Resources: Bureau of Economic Geology, The University of Texas at Austin, prepared for Texas House of Representatives, Joint Interim Committee on Energy, 39 p.

Jones, P. H. (1976) Natural Gas Resources of the Geopressured Zones in the Northern Gulf of Mexico Basin: in Forum on Potential Resources of Natural Gas: Department of Geology, Louisiana State University.

Loucks, R. G. (1978) Geothermal Resources, Vicksburg
      Formation, Texas Gulf coast: Draft Report to the U.S.
      Department of Energy, Bureau of Economic Geology, 16 p.

Papadopulos, S. S., Wallace, R. H., Wesselman, J. B., and
      Taylor, R. E. (1975) Assessment of Onshore Geopressured-
      Geothermal Resources in the Northern Gulf of Mexico
      Basin: U. S. Geological Survey Circular 726, p. 125-
      146.

WALL STREET JOURNAL (April 27, 1977) 1,001 Years of Natural
      Gas, Editorial.

WALL STREET JOURNAL (May 20, 1977) ERDAgate!, Editorial.

WALL STREET JOURNAL (May 27, 1977) The "Energy Crisis"
      Explained, Editorial.

6

Nuclear Energy Resources

Leon T. Silver

The interactions of the geological sciences with nuclear energy power development have been continuous since the first uranium fuel supplies were sought but they have recently intensified at an even faster pace than the growth of this remarkable energy source.

From the initial experiments in power production 30 years ago, the nuclear electric power industry has grown to comprise about 65 active light water reactors (LWR) which produced about 12 percent of the electricity generated in 1977 (1). If the contemplated three-fold expansion of the number of reactors (to about 207) is achieved by 1990, nuclear power will probably supply 25-30 percent of the growing national electricity requirement, according to recent ERDA analysis (2). The existing reactors, and those currently planned, along with supporting facilities and fuel requirements, will represent an investment of several hundred billion dollars, approaching in magnitude the United States national debt.

The longer term growth pattern of national nuclear power capability is uncertain. As a relatively youthful power development with enormous potential, nuclear energy has appeared at a time when traditional hydrocarbon and coal fuels are recognized to be finite resources having unique value in alternative uses. Nevertheless, the design of the long-term role of nuclear electrical energy in the United States energy plan has not been completed. There does not now exist an unequivocal mandate for the magnitude and forms of its utilization in our long-range energy development program.

At this point in time, national planners in many other leading industrialized countries have committed their economies to nuclear power generation as a major, if not dominant, replacement system for fossil fuels. In this country and elsewhere, an ongoing debate centers around the suitable role of the nuclear power option, focusing particularly upon concerns for adequate controls in handling reactor grade fuels and reactor spent fuel products. The social and political aspects of the debate process seem to be contributing a restraining influence on the current perceptions and planning of the United States electric utility industry with respect to further investments in nuclear power. Technical and economic questions involving fuel supply, regulatory environment, rate controls, and taxation also loom prominently for the industry. Until some clear and comprehensive national decisions emerge, the prospects for future utilization of nuclear energy in the United States will be difficult to assess. Nevertheless, a valuable and productive nuclear power industry exists today.

The geological sciences have been requested to provide a broad variety of inputs to the operating nuclear industry, to national planners and to the public forums considering nuclear energy policies.

When the industry was in its infancy, it could be said to have been conceived by physicists and delivered by chemists and engineers. To support the growing industry, geological scientists have been asked to find uranium to feed fuel to the growing number of reactors, help bed them in suitable sites, and to assist in planning containment and disposal of wastes.

In the active discussions concerning the nation's future nuclear utilization, earth scientists are called upon to assess the longer term availability of uranium and thorium relative to the availability of various competitive fuel resources. They are asked to provide suitable geological criteria and specific site nominations for the effective management of radioactive wastes over intervals of time of geological dimensions, from thousands to a million years, or more. Frequently, they are requested to provide with these complex assessments and recommendations, statements of precise numerical confidence values suitable for confident planning. Geological professionals strive to respond responsibly to such requests, of course, but they take some care to emphasize the information limits and the extent to which subjective analysis enter, of necessity, into such important considerations.

Nuclear Energy Resources

The light water reactor design which predominates in the U.S. nuclear industry requires that the ^{235}U isotope be enriched from its abundance ratio of 0.7 percent in natural

uranium to several percent in fuel. The other 99.3 percent,
^{238}U, is not effectively fissile in the LWR until it is converted to ^{239}Pu during reactor operation. Present executive
policy is to not reprocess spent fuel to extract the ^{239}Pu
for reactor use. With this policy, and with the expected extraction efficiency for ^{235}U during natural uranium processing, it appears that about 12,000 short tons (ST) of U_3O_8
were required in 1977 to fuel the active reactor systems. By
1990, the annual requirements for the planned reactor systems
under similar operating assumptions, will be between 45,000
and 60,000 ST U_3O_8 and the cumulative requirements between
400,000 and 500,000 ST U_3O_8. Depending upon the continuing
rate of growth of the nuclear power program, the capacities
and efficiencies of ^{235}U enrichment systems, and continued
limitations on spent fuel reprocessing, estimates of the cumulative uranium requirements for the year 2000 range from
about 1,000,000 to more than 1,3000,000 ST U_3O_8(3,4).

In thirty years, through 1977, the United States produced
slightly more than 310,000 ST U_3O_8. Past annual production
is shown in figure 1. Only a small fraction of this production is in available inventory. In the next 23 years, a total production 3 or 4 times greater must be achieved to meet
even conservative estimates of demand. Figure 2 shows the
projection of uranium production from operating centers and
the probable production from new mines and mills now being
activated, compared to two estimates of demands. It is apparent that remarkable new annual production levels must be
achieved within 10 years to support the power plants operating or planned for operation in the next twenty-five years.

What are the resources to support this nuclear economy?
Some pragmatic definitions are needed to discuss this question. Geologists and mining engineers distinguish between
reserves (profitably extractable concentrations of metal in
ore, positively identified as to location, quantity and grade
by adequate three-dimensional sampling) and other forms of
resources which are only partly defined or subeconomic, or
whose existence is anticipated on the basis of geologic models but has not yet been identified or confirmed. The latter
resources are largely hypothetical or speculative, and sometimes are called potential resources. The significant differ-
ence between reserves and potential resources is the identification and confirmation process by which reserves are established. Potential resources remain to be discovered and
tested. It is only on the basis of proved reserves that development funds are raised, mines and mills are built, and
that electrical utilities can establish assured fuel supplies
on which to base plans for new nuclear power plant investments.

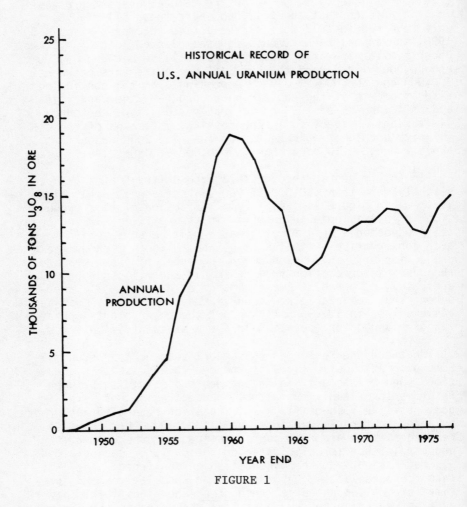

FIGURE 1

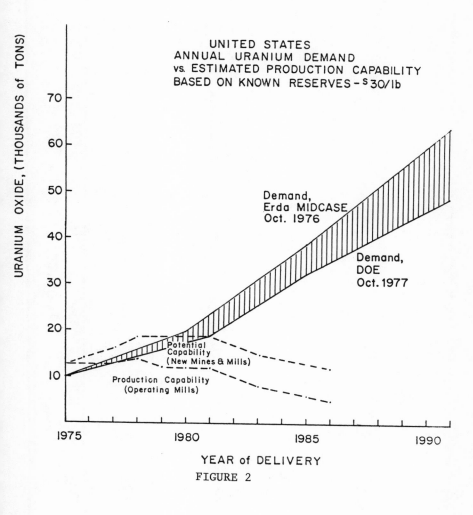

UNITED STATES
ANNUAL URANIUM DEMAND
vs. ESTIMATED PRODUCTION CAPABILITY
BASED ON KNOWN RESERVES – $30/lb

Demand,
Erda MIDCASE
Oct. 1976

Demand,
DOE
Oct. 1977

Potential
Capability
(New Mines & Mills)

Production Capability
(Operating Mills)

YEAR of DELIVERY

FIGURE 2

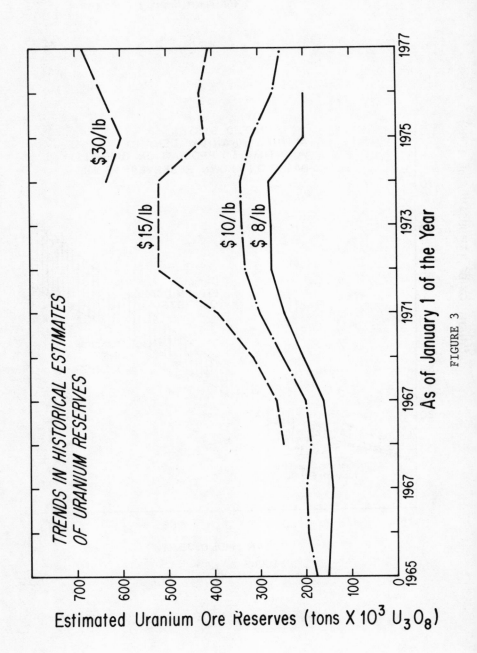

TRENDS IN HISTORICAL ESTIMATES
OF URANIUM RESERVES

$30/lb

$15/lb

$10/lb

$ 8/lb

Estimated Uranium Ore Reserves (tons X 10³ U₃O₈)

700
600
500
400
300
200
100
0

1965   1967   1967   1971   1973   1975   1977

As of January 1 of the Year

FIGURE 3

According to detailed anlyses by the Department of Energy (4,5,6), reserves of between 650,000 and 700,000 ST U_3O_8 which could be produced as ore at a remaining production cost (forward cost) of $30 per pound of U_3O_8, exist in the United States. This figure includes all lower forward cost categories. DOE sources imply the uncertainty in this reserves figure is somewhat greater than $\pm$ 20 percent (5). Historically (in figure 3), this value has remained constant within that uncertainty for the last few years. Similarly, trends of reserve estimates for higher grade ores, less expensive to produce, have not climbed in recent years. An important reason is that only the higher grades of ore have been economically producible until recently. The 1976 average price is reported at $16.10 per pound U_3O_8 (7).

Although new additions to all categories of reserves have not significantly surpassed the rate of annual production, this has not been for lack of vigorous effort. The uranium exploration industry, spurred by an increasingly favorable market, is drastically increasing its expenditures, manpower and acreage explored. The most practical index of this renewed intensity of the exploration effort is the annual drilling footage expended to explore for and develop uranium deposits. In figure 4, the historical pattern of exploration drilling as well as total drilling is shown and projected through 1978 (3). Because the cost of drilling has risen far more rapidly than inflation, reflecting greater average depth of exploration and more sophisticated techniques, annual exploration expenditures have been growing at even greater rates. Total exploration expenditures in 1971 were 41.3 million dollars; in 1974, 79.1 million dollars; in 1977, an estimated 236 million dollars. Although the yields of U_3O_8 reserves per foot of drilling and per dollar expended are continually declining, we can plausibly expect some improvement in annual discovery rates. Will it be sufficient to meet the projected demand?

The Uranium Resource Base

Many analysts have employed estimates of the total uranium resource base (reserves plus potential resources) to develop answers to this question of sufficiency. DOE has offered a recent preliminary estimate of about 3.5 million ST U_3O_8 producible at forward cost of $30/lb. (6). Less than 20 percent of this $30/lb. uranium resource is in identified form, as reserves. No statement of uncertainty for the overall figure has been offered but it must be considerably larger than the $\pm$ 20 percent which ERDA suggested for its $15/lb. forward cost reserves (5).

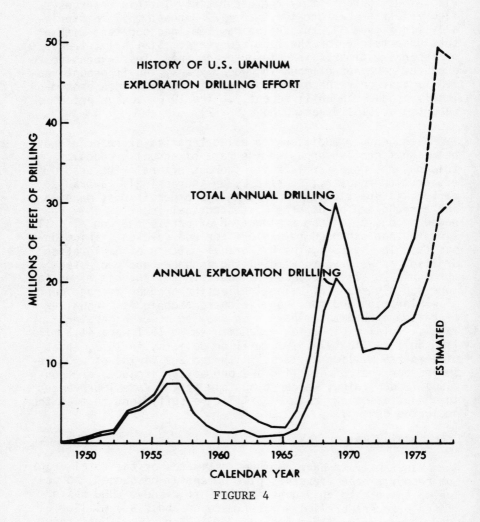

FIGURE 4

There is broad recognition that the national uranium base cannot be assessed accurately at this time. Nearly 90 percent of our reserves and much of our implied resources are located in certain sandstone host environments in the Colorado Plateau-Rocky Mountain and Gulf Coast regions. Estimates of those resources which are believed to be extensions of known ore deposits and known mineralization trends in productive districts, based on geologically reasonable models for their formation, are more persuasive than inferences concerning other classes of potential resources. These most confident resources are included in a category, probable potential resources, by DOE - ERDA which has estimated them at nearly one million tons at $30/lb. forward cost in the sandstone occurrences and at 1.09 million tons for all types *[see table 1 (6)].* Most of this probable potential resource must yet be discovered and converted to reserves from which production can then be planned.

DOE also estimates another 1.6 million ST U_3O_8 ($30/lb.) of possible or speculative potential resources may exist in districts or geological provinces where there has been no history of significant previous uranium production. Again, almost all of this material remains to be discovered and tested. The uncertainties in their numerical estimate for this segment of potential resources have not been stated by DOE-ERDA.

A significant increment of potential uranium supply may be expected as a by-product from phosphate (and, to a lesser extent, copper) mining. Central Florida phosphate production may yield up to several thousand ST U_3O_8 per year, if extraction technology is effective and sustained phosphate and uranium markets are favorable. DOE expects more than 150,000 ST U_3O_8 from this source in the next 30 years (4), assuming significant production starts soon. That high expectation has been expressed for several years but in 1977 annual by-product uranium was about 100 ST U_3O_8. It will be 1980 or later before practical levels of sustained uranium by-production can be determined.

The Energy Research and Development Administration initiated the National Uranium Resource Evaluation (NURE) program in 1974 to achieve a more effective assessment of the national uranium potential. It is scheduled for completion in 1981, and includes aerial radiometric reconnaissance; subface geologic and related investigations; hydrogeochemical and stream sediment reconnaissance; subsurface geologic investigations; and technology development to improve existing methods and design new equipment for use in all phases of uranium exploration. The U.S. Geological Survey is

Table 1

(ERDA)

U.S. Uranium Resources

January 1, 1977

Tons U_3O_8

$/Lb. U_3O_8 Cost Category	Reserves	Potential Resources [1]		
		Probable	Possible	Speculative
$10	250,000	275,000	115,000	100,000
$10-$15 Increment	160,000	310,000	375,000	90,000
$15	410,000	585,000	490,000	190,000
$15-$30 Increment	270,000	505,000	630,000	290,000
$30	680,000	1,090,000	1,120,000	480,000
$30-$50 Increment	160,000	280,000	300,000	60,000
$50	840,000	1,370,000	1,420,000	540,000

[1] The reliability of the potential resource estimates decreases from the probable to the speculative class.

assuming a major role in developing many important geological
and geochemical aspects of the entire assessment program.
Through rapid and effective information and technology transfer to industry, the NURE program hopes to contribute to increased industry exploration success. The dimensions of federal funding for the NURE program are reported to be expanding significantly as the implications of a possible shortfall
in uranium within the next ten years have become apparent.

The economic limitation imposed upon the definition of
the useful resource base at any given time is the minimum
price which will support production from a marginal uranium
deposit. This figure will change with demand. It has been
suggested that the nuclear industry can support higher fuel
prices because nuclear power costs are not as sensitive to
fuel prices as are costs in other power-generating systems.

It is only in the last few years that a percentage of sales
in the range of $30 to $60 have made production of much of
the $15-30/lb. forward cost increment of reserves appear
profitable. How would higher market prices affect the resource base?

For resources dominated by sandstone-type uranium deposits, it appears that the magnitude will not increase in direct proportion to price, much less exponentially. In figure
3, it is clear that the $30/lb. reserve category in 1977 rep-
resents only a 65 percent increase over the $15/lb. category
which it includes. This expansion with a doubling of cost
category is somewhat exaggerated by the past depletion by
production of the $15/lb. reserves. When ERDA recently re-
leased its first estimate of $50/lb. reserves (table 1), it
expanded the identified reserves from 680,000 tons, $30/lb.,
to 840,000 ST U_3O_8 at the higher cost, a 24 percent estimated
increase in uranium in response to a 67 percent cost change.
Several factors involving information availability may have
depressed, somewhat, the true tonnage to price relationship.
Nevertheless, there appears to be substantial mining and geo-
olgic data to indicate that for presently-identified types of
resources, drastic price increases will be required to substantially influence the total resource figures.

At this stage in our understanding of our national uranium resources, and with current economics, it is prudent to
recognize that the only resources for which a quantitative
assessment can be made which may be available for production
to the year 2000, are the reserves (680,000 ST), the probable
potential resources (1,090,000 ST), and perhaps 60,000-70,000
tons of by-production from phosphate mines. These sum to
about 1.83 million ST U_3O_8, at $30/lb. This figure would

Table 2

Potential Resources to Production:
Activities and Time Requirements

Activity	Typical Time Requirements
Land Acquisition and Exploration, including permits............................	5 – 10 years
Property Development, including permits............................	1 – 3 years
Mine and Mill Construction, including permits............................	2 – 3 years
Total Time from initiation of exploration to beginning of production............................	8 – 16 years
Typical Production Life for active mines in 1976............................	16 years

appear capable of meeting our presently-perceived needs to
the year 2000, with two important caveats: (1) the uncertainty in the sum may be more than one-third of the total
figure; and (2) that more than half this sum remains to be
discovered and verified long before then.

Uranium Supply Problems, 1980-2000

Uncertainties (and controversies) about the total uranium resource base have tended to obscure important factors of
more critical concern for the intermediate period 1980-2000,
then production from current reserves must be supplemented
and replaced by production from what are now potential resources (figure 2).

Although the precise time for inception of this new increment of production is uncertain, it should occur between
1980 and 1985, perhaps about 1982. The U.S. government perhaps can provide an unspecified ^{235}U-enriched inventory for
an interim period of one or two years. However, the U.S. also faces international commitments to supply enriched fuels
to foreign countries cooperating with our spent fuel non-
reprocessing program. There is increasing evidence suggesting that for the entire western world, the probable uranium
demand and production curves will cross to a shortfall relation, between 1982 and 1990, without an accelerated exploration effort (8).

It seems clear that the United States must start convert-
ing its potential resources to reserves at a much more rapid
rate than has been recorded recently. The category, probable
potential resources, should and will receive greatest attention for near-term exploration efforts. Simultaneously, however, exploration in all classes of favorable geologic prov-
inces should be pressed to establish new resources and to dis-
cover important new districts.

The conversion of potential resources to production consists of a number of essential sequential operations, each of
which requires significant time for completion (table 2).

It is apparent from these lead times that discoveries
and reserves developed in the next five to ten years will determine the production level from 1990 to 2000, and that it
is the current exploration success rate that will influence
supply in the mid 1980's. Since mine production is distributed over a mine lifetime of 10 to 20 years, one can expect
only 1000 ST U_3O_8 of annual production over a ten-year period
from a hypothetical 10,000 ST U_3O_8 discovery, starting no
earlier than five years after the discovery. To obtain

40,000 tons of new production capacity in 1990, the exploration industry must achieve at least 400,000 ST U_3O_8 of discovery and confirmation of new reserves before 1985. This corresponds to an annual level of success rarely sustained in the past, but it does not appear impossible. The challenge for exploration geologists beyond 1985 will continue to grow, unless major changes occur in the boundary conditions governing national use of nuclear electric power generation.

In our earlier exploration experience the two largest additions to reserves were directly associated with the discovery and development of two rich regions, the southern San Juan basin, New Mexico, and the Tertiary basins of Wyoming. For the future, it appears that similarly large but entirely new districts must be found at even more frequent intervals.

The overall success of the national exploration efforts will depend on (1) the character and dimensions of the resource base, (2) the incentives for industry to invest in accelerated exploration, and (3) the scientific and technical qualities of the exploration programs. The first is out of our hands. The second will be determined by long-overdue national policy decisions and implementation efforts. The third will be the responsibility of geologists, geochemists and geophysicists.

Industry already has organized impressive exploration teams; many of the best are drawn from the petroleum companies with their great exploration expertise in sedimentary basins. The federal government's field and laboratory programs, led by the U.S. Geological Survey and DOE contractors, are focusing their technical talents on the NURE assessment program. Perhaps the most technically challenging problem is the rapid development of improved and more diverse generic models for the formation of uranium ore deposits. The exploration industry, itself, frequently has expressed the need for expanded basic research in this area.

Very little national progress has been made in the geology and geochemistry of uranium deposits in the last 10-15 years. This deficiency can be attributed, in part, to a period in the 1960's when the USAEC, faced with an apparent uranium surplus, withdrew its university support for basic research in the geology and geochemistry of uranium. Many outstanding scientists in universities who were trained in laboratories built with AEC support, turned to other activities. (Not incidentally, many of them made important contributions to the U.S. geodynamics and to the NASA lunar and planetary exploration programs).

The Department of Energy can and must now enlist a broad spectrum of university researchers (including many of these same AEC products), with their new concepts and tools to participate in the basic research and technique development required for more creative and constructive approaches to uranium discovery. It appears to the writer that the international challenge for scientific leadership in this area is severe.

Earth scientists are now deeply involved with the rest of the science community in addressing the technical problems of successful utilization of nuclear electrical energy. In this analysis it has been emphasized that the pace of technical efforts in broad areas of resource assessment, exploration, and mine and mill development must be drastically increased by factors of 2 to 10 to meet the near and intermediate term uranium requirements. For the longer term, it appears that for the most effective use of nuclear power, national planners should be preparing a program of reactor designs and construction in which fuel-breeding from infertile ^{238}U and ^{232}Th will amplify the domestic uranium resources. Solutions leading to maximum efficient utilization of natural fuels will not reduce the many diverse technical roles which geologists, geochemists and geophysicists must play if nuclear electrical power is to become a dominant energy source in our society.

References

(1) Federal Energy Administration, "Monthly Energy Review," August, 1977.

(2) ERDA - GJO - 108(76), "Uranium Industry Seminar," October, 1976.

(3) ERDA - GJO - 100(77), "Statistical Data of the Uranium Industry," October, 1977.

(4) DOE - ERDA - GJO - 108(77), "Uranium Industry Seminar," October, 1977.

(5) ERDA - GJO - 111(76), "National Uranium Resource Evaluation, Preliminary Report," July, 1976.

(6) ERDA - GJO News Release, June 22, 1977.

(7) ERDA - (77) - 46, "Survey of United States Uranium Marketing Activity", May, 1977.

(8) NEA (OECD) - IAEA, quoted in "Nuclear Fuel", Feb. 20, 1978.

Energy, Environment, and the Geosciences

M. Gordon Wolman

Increasing demand for energy, the need to develop alternatives to heavy reliance on oil and gas, and recognition that the choice of energy source, technology, location, and conversion all have significant environmental implications, have fueled the energy-environment debate. An extreme view, which in all likelihood no one really believes, holds that the conflict is an either-or one which of course cannot be true for mankind. Assuming anything like present population levels on the earth, a great deal of energy will be needed even if the technological aspirations of the future society are modest. On the other hand, progressive deterioration of man's environment on earth is antithetical to any future for mankind and society.

Because a great deal has already been written about energy and environment, four points are chosen here for emphasis: 1) the unique custodial role of the earth sciences in relating the concepts of time and space to public policy, 2) the requirement that assessment of environmental impacts of alternative energy policies include the impacts associated with extraction, processing, transportation, conversion, use, and disposal of wastes and not simply extraction, 3) the demands upon knowledge of the geosciences posed by decision matrices including each of the above stages along with combinations of alternative technologies, locations, and environmental objectives applicable at each stage, and 4) the importance to policy decisions of enhancing knowledge of some basic phenomena in the earth sciences illustrated by problems posed in predicting environmental effects of mining and processing oil shale, burning of fossil fuels, and storing of nuclear wastes.

Geosciences: Custodians of Time and Space

The contribution of the earth sciences to the continu-

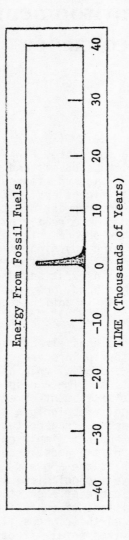

Figure 1: "Human Affairs in Time Perspective" by Hubbert, 1949.

ing formulation of interrelated policies called energy and environment is, I believe, both philosophic as well as strictly scientific or technological as the latter words are customarily used. Earth scientists (and I include the astronomers) are unique among those engaged in the process of inquiry and policy formulation--not simply because of their knowledge and skills--but because they are the custodians of the perspective of time and scale. Without this perspective no proper environmental impact statement can be written, nor can society formulate energy policies aimed at cooking the eggs without killing the golden goose that lays them.

It is perhaps not an accident that the person who first jarred Americans' consciousness into recognition that oil and gas are "finite" resources was a geologist. Perhaps only a geologist would drive home the message with a graph (figure 1) showing a single small blip with a span of about 100 years representing the total oil and gas use by man on a time scale of 40 to 80 thousand years (Hubbert, 1949). (Then too, perhaps only King Hubbert would have done both of these things!) The lesson in time and scale, however, is useful. It is also interesting to note that it is almost thirty years since publication of Hubbert's diagram, a span roughly one-sixth of the useful life of the resource, and policies to cope with the problem are only beginning to be formulated.

Earth scientists become the custodians of scale because the earth and its processes define the boundaries of environmental problems. Some energy-environment problems are global, others regional, others local. Because the earth is divided into major geologic and climatic regions having unique assemblages of properties, these spatial units help both to identify energy options and to define the kinds of environmental effects which one can be expected from specific energy developments and uses. Active faults around the Pacific, an air pollution disaster at Donora, Pennsylvania, increasing salinity in the Colorado River, and the distribution of potential sites for disposal of high level nuclear wastes are not random spatial phenomena. The significance of earth scales in space and time recurs repeatedly in considering the interaction of the demands for energy and environmental management.

Three observations about the earth essential to understanding the environmental implications of human actions stem from the domain of the earth sciences: the fact that the earth is finite (even if the imagination of man is not), that not only the earth itself, but the land, air, and water are in constant motion, and that change and variation rather than stability characterize many features of the earth's

surface have by now become hackneyed truths to which every-
one subscribes whether fully understanding their implications
or not. Only tens of thousands of years ago the oceans were
one hundred feet lower than they are today, mean global
temperature was 10°F to 15°F degrees colder, and forests and
animals quite different in most parts of the globe than they
are today. Earth processes are also highly variable over
time. Because news media follow the progress of bulges in
the "solid earth" of southern California, report the magnitude of earthquakes around the world on the Richter scale,
and daily note devastation by flood, drought, wind and wave,
the public at least hears about the variability of nature.
It is less clear whether the public and its decision makers
are fully aware of the implications of this variability on
the ability to predict the environmental impacts of particular energy policies, or of the effect of variability on estimates of the magnitude of the risks involved in tampering
with the earth in certain places at certain times. Earth
scientists may take some credit for what appears to be increasing awareness of the dynamics of nature, although many
would contend that the environmental movement is responsible
for the new consciousness. More important now is that this
awareness contribute to enlarging the scientific base on
which reconciliation of conflicting energy and environmental
demands depends, and to enchancing the rationality of the
process of decision making.

Energy-Environment Matrix

Despite the attention it has received, a discussion
of the relationship of the geosciences to energy development
and the environment requires reiteration of the "holism" required in evaluating the environmental impacts of alternative policies. (Impact here is used in the sense of effect.
Unfortunately impact has come to mean either effect, or
both effect and value.) The environmental impacts of alternative energy options must encompass extraction, processing,
transportation, conversion, use, and disposal of wastes.
Further, each potential energy source which is itself or
requires uses of raw materials (wind, solar) has a specific
composition (the sulfur content or BTU quality of coal), and
a geographic location. In addition, the location of each
step may also vary as well as alternative technologies used
at each stage. This matrix of options and their environmental effects has been pioneered by Bower and others (1977)
and is currently being used extensively in evaluating environmental impacts of potential energy developments. A few
such studies are used illustratively eleswhere in this paper.

Each step or activity generates a set of impacts which might be mitigated by alternatives including changes in location or in the technology of a process or the treatment of wastes. The impacts of each may occur on the land, air, and surface or subsurface waters and the biota associated with each, and adequate assessment of such impacts involves knowledge of sources, pathways, intervening reactions, and sinks for each step from extraction through disposal of residuals. (I omit the biota including man. These interactions, ultimately the most important, cannot be derived without a knowledge of the scale and dynamic behavior of the natural system. A tabular matrix (table 1) illustrates the differential environmental impacts of the generation of 1,000 megawatts of electricity from coal mined from a deep or surface mine with conversion at the mine mouth or at the load center, assuming three possible levels of environmental control. Even this simple comparison generates a matrix consisting of ninety boxes.

Obviously a matrix comparing the environmental effects of many alternative energy sources, location, attributes, and technologies would be enormous. Whether explicit or not, consideration of energy-environment alternatives involves evaluating the cells in such a matrix, as well as placing values upon each. Each entry in turn requires a piece of information about earth materials and processes. Some information, such as land disturbed per ton of coal mined may be relatively well known for a given level of technology. In contrast, the ability to predict acid, dissolved solids, or sediment generated, or subsidence produced, per ton of coal mined must rest not only on observations from previous experience but on a sophisticated knowledge of hydrologic and geologic processes. The degree of accuracy required in prediction is dictated by the importance of the information in the decision. Some information, as noted below, may be critical to informed decision.

The task of defining regional and larger environmental effects is nicely reflected in the atmospheric "transport and fate" studies reported by the interagency research group in the Second National Conference held in 1977 entitled Energy/Environment II, sponsored by U.S. EPA (1977). This work includes studies of Midwest interstate sulfur transformation and transport, multi-state atmospheric power plants, regional transport of sulfur, and transport from western power plants and its effects on cloud precipitation. Like the studies of acid rain in New England (Likens, 1972) and in Europe and Scandinavia (Ottar, 1975), such inquiries define the regions over which a set of impacts will be felt as well as the interactions among manmade activities and natural pro-

Table 1 - RESIDUALS MANAGEMENT COSTS AND RESIDUALS DISCHARGES, THREE LEVELS OF ENVIRONMENTAL QUALITY: COAL-ELECTRIC ENERGY SYSTEM

	Case I			Case II		
	Mine-mouth plant high-impurity seam, area-strip mine			Load-center plant, low-impurity seam, deep mine		
Environmental Quality Design Level	minimal I	moderately high II	high III	minimal I	moderately high II	high III
Price of power:	mills/kwh			mills/kwh		
At busbar	6.00	7.05	8.57	6.94	7.92	9.37
At load-center substation	7.64	9.00	11.78	7.17	8.20	9.98
RESIDUALS FLOW TO ENVIRONMENT (Annual basis, except for transmission)						
Acres of disturbed land/Acres of reclaimed land	980/0	1040/1040	1250/1250	not applicable		
Gross increase in mine drainage:						
million gallons	not applicable			110	80	90
tons of sulfuric acid	not applicable			360	180	0
Preparation of plant water (1 to 8% solids), million gallons	0	0	0	0	0	0
Preparation plant refuse, tons	0	1,600,000	2,600,000	0	0	300,000
Preparation plant airborne dust, tons	0	4,400	900	0	0	700
Power-plant stack emissions, tons:						
Particulates	140,000	900	100	25,000	700	100

Table 1 (continued)

	Case I			Case II		
	Mine-mouth plant high-impurity seam, area-strip mine			Load-center plant, low-impurity seam, deep mine		
	minimal	moderately high	high	minimal	moderately high	high
	I	II	III	I	II	III
Sulfer oxides (sulfer content)	200,000	29,000	15,000	28,000	7,500	3,500
Nitrogen oxides (nitrogen content)	10,000	8,000	2,000	10,000	8,000	2,000
Power-plant solid waste, tons	1,000,000	950,000	800,000	300,000	450,000	300,000
Thermal discharge to watercourse, billion Btu	6,000	0	0	6,000	0	0
Water consumption (extra evaporation), million gallons	7,500	5,000	0	7,500	5,000	0
Transmission-line towers:						
lattice type	4,000	2,500	0	120	0	0
tubular poles	0	2,000	4,000	0	150	0
underground circuit-miles/Total circuit-miles	0/800	0/880	80/800	0/25	0/28	25/25

Description of Table 1

Residuals management costs and residuals discharges, three levels of environmental quality: coal-electric energy system (from Delson and Frankel, 1972). A "simple" matrix including 90 cells each of which requires information in the earth sciences to predict potential environmental effects.

cesses. The policy implications of such studies, the controversy over tall stacks or "scrubbers" or mine-mouth generation
of electric power in western coal fields versus coal transport
are clear. Thus one might well conclude, in view of the
thickness of the coal seams and relatively thin overburden,
and the available though limited moisture that the physical
and biological environment of the Powder River Basin in
eastern Wyoming could better withstand the effect of twenty
to twenty-five coal mines than could more widely scattered
locations dotted by large coal mines operating in less favorable geologic and climatic settings throughout the West.
However, one might view energy conversion within the region
less sanguinely if studies demonstrated significant potential
for regional deterioration in view of the importance of clean
air to the region's other activities. Thus the spatial scale
of the problem is essential in weighing the potential advan-
tages and disadvantages of mining and shipping low sulfur
coal to be converted at the cities in the midwest or con-
verting at the mine and shipping electric power. Regulatory
strategies at all governmental levels are dependent upon
this same knowledge of relevant spatial scales.

A detailed inquiry into the present state of
predicting environmental effects within the sphere of the
geosciences for all energy sources from extractions through
conversion and transmission would try the patience of the
reader and the capacity of the writer. Experience suggests,
however, that the available knowledge is quite limited, particularly as one moves from broad generalizations to the prediction of earth processes and environmental effects of new
and major human activities in poorly studied regions. As
noted in the conclusions, however, these facts, rather than
paralyzing action, simply commend inquiry as well as skepticism in evaluating information and restraint in uninformed
judgement.

Uncertainty in Some Areas of Fundamental Importance

Three areas of uncertainty in the earth sciences for
which information is of fundamental importance in assessing
environmental problems associated with energy development are
treated very briefly here to illustrate the close tie between
policy formulation and our knowledge of the earth.

Extraction and Refining:
Surface Mining and the Hydrologic Regimen

The four-year battle over surface mining legislation
focused attention on past horrors and the real potential for
future ones in indiscriminate gouging of the land surface

without reclamation. It also produced evidence that destruction need not accompany extraction of coal or other minerals if the work was done properly and in the right place. But complex geologic and hydrologic problems remain. Further, the problems of mining cannot be separated from conversion and consumption of water. Depletion of flow and addition of salt loads combine to increase salinity (Howe, 1977, James and Steele, 1977).

Prospective plans for mining and retorting of oil shale in western Colorado, Utah, and Wyoming, for example, pose questions which remain only partially answered.

In summarizing present information Myers et al of EPA (1977) estimate that each barrel of kerogen (oil product) produced will use 2 to 6 barrels of water, 1 to 2 tons of shale, and 100 square feet of land, resulting in the production of 1/3 lb of airborne dust, 2 1/2 lb of polluting gases, 2 to 5 gallons of contaminated water, and 1 to 1 1/2 tons of spent shale. A measure of the magnitude of the endeavor is given by the fact that the proposed development calls for mining of 61,000 tons of oil shale per day for twenty years.

Spent shale contains a variety of salts and the exposed land areas in which it is placed would be subject to runoff from intense storms. Should the salts escape in surface waters, or be leached by water moving through the ground the potential exists for contaminating limited supplies of fresh water in the tributaries of the Colorado River. Salt concentration in the Colorado at the Mexican border is currently 890 ppm, sufficiently high to require consideration of desalinization of Colorado River water for delivery to Mexico under the terms of a recent treaty. Decades ago prior to major irrigation development salinity of the Colorado was 238 ppm (Wolman, 1971). Projections of present trends indicate that by the year 2000 salinity in the river could reach 1100 ppm. Calculations by Howe (1977, p. 133-136) incorporating measures of economic activity and water use in the region suggest that reduction of marginal irrigated acreage might be cheaper than desalinization in reducing total dissolved loads in the river. Work remains to be done on trade-offs between various water-using activities including agriculture, oil shale development and coal gasification and their effect on the quantity and quality of available water supplies. The behavior of seepage water in irrigated agricultural as well as in shale and coal development are related to overall energy development and its effect on environmental quality.

Difficult problems also confront the analyst in esti-

mating the probable movement of sediments and flow from both
disturbed and natural areas in semi-arid regions. Alternating
cycles of cutting and filling of gullies or arroyos with sediment are well documented in the western United States. Climatic change, large floods, variation in rainfall characteristics, overgrazing by sheep and cattle, and construction activities by man have all been implicated in these cycles
(Cooke and Reeves, 1976). In all likelihood, all have played
a part in the process at some time and in some place. Yet
records of flow and sediment movement are short, and understanding of the processes of discontinuous transport of sediments on the land surface and in channel is limited. Site
specific estimates of these processes in local areas may
range from poor to good. However, the immense variability of
this "natural" background places great uncertainty on the
estimates of probable rates of erosion, sediment movement
and disposition over long periods of time, and in turn upon
the storage required to detain large volumes of wastes.

Subtle changes in the flow of ground and surface waters
from disturbed lands may alter the conditions of flow in
ephemeral channels in semi-arid regions. Where water is
scarce, meadows in bottomlands provide a major source of
grassland for grazing and hay production for livestock. The
limited and variable flow, characterize the importance of the
resource while contributing to the difficulty of predicting
its behavior. Alterations of the landscape are likely to
modify the hydrologic regimen but it remains difficult to
predict precisely what changes will occur, an essential step
in predicting the negative or positive impact of a change.

The focus on energy resources in semi-arid regions
coupled with concern about the management of water for all
uses should stimulate inquiry into the behavior of water,
sediment, and other constituents to strengthen confidence in
the elaborate models used in arriving at policy choices.

Conversion: Climatic Change and Burning of Fossil Fuels

A recent evaluation of the potential for inducing
climatic change by the burning of fossils by a committee of
the National Academy of Sciences (1977) has prompted a sharp
increase in research into the carbon dioxide cycle and the
role of CO_2 in controlling climate. In a carefully worded
statement the committee concluded that, "It has become increasingly apparent in recent years that human capacity to
perturb inadvertently the global environment has outstripped
our ability to anticipate the nature and extent of the
impact." Noting that, "the results of their study should
lead neither to panic nor to complacency," the committee spoke

of a "lively sense of urgency" in moving forward to "illumin-
ate the scientific uncertainties" in order to place "credible
information" in the hands of policy makers in their search
for long term energy policies.

The conclusions of the NAS committee are based on the
observation that while different investigators arrive at
somewhat different estimates, all agree on a probable four to
eight-fold increase in atmospheric carbon dioxide within
roughly the next 100 years resulting in an estimated increase
in average world temperature of 6° C. or more, with increases
at the poles of three times this amount (NAS, 1977, foreword
p. 2). This fluctuation far exceeds those of the past
several thousand years, and could also have a major impact
on global precipitation and on sea level with accelerated
melting of glacier ice. The potential impact on human activities is perhaps incalculable. Urgency is demanded not by
the immediacy of a political decision, but because the information required to evaluate these prospects will take decades
to acquire.

The effort requires information on the movement and
distribution of CO_2 in the atmosphere, biosphere, and oceans,
on the relation of CO_2 to atmospheric and climatic processes
(the dynamics of the atmosphere), and on the ways in which
the activities of man alter the disposition and behavior of
the carbon dioxide cycle. It is interesting to note that the
fundamental work on the carbon dioxide cycle, on climatic
variations, on carbon-14 (an isotope of carbon), on oxygen
16/18 ratios, and on layering in glacial ice used in recon-
structing past temperatures all began before the recent surge
of interest in the environment. The work was motivated by a
fascinating diversity of interests in the history and dynamic
behavior of the earth. Happily, the prompting of the committee has been heard and research has accelerated on this
fundamental scientific, long range yet socially relevant
problem.

Waste Disposal: Nuclear Power

Despite a slower growth in construction of commercial
nuclear power plants than some envisioned a decade ago
(O'Leary in U.S. EPA, 1977), interest in locating sites for
nuclear wastes from commercial reactors has burgeoned, in
part because of moratoria on plant operation in some states
until the locations of such facilities are determined. The
quantity of such wastes projected for the next several decades remains small relative to the quantity and variety of
wastes already accumulated in military operations and
awaiting "permanent" disposal. Both require establishing

site suitability criteria to be followed in the process of
site selection. Evaluating the criteria and assessing alternative kinds of sites, whether these be on land or beneath
the ocean, requires detailed knowledge of the geology at
particular locations within extended regions, coupled with
sophisticated knowledge of hydrologic and geologic processes.
There is perhaps no better illustration of the need for the
perspective on time and space of the earth scientist.

From a geological perspective the problem of containment, assuring that radioactive materials do not escape to
the surrounding environment over specified periods of years,
differs significantly depending upon whether one is interested in designing a facility from which stored wastes might
be retrieved within a "brief" period (geologically speaking)
of decades or hundreds of years, or designing a permanent
containment for thousands of years for long lived level
high radioactive wastes. In many ways, however, from a
social perspective decision makers today are likely to look
for minimum movement of radioactive materials for any site
designated as "long term," whether measured in generations of
human beings or geological eras. The problem thus posed con-
fronts real unknowns. The kinds of geological formations
sought are ones which traditionally in the field of flow in
permeable materials would generally be characterized as
"impermeable" - yet we are now interested in their permeability or conductivity - that is, the possibility that
migration of fluids will actually occur even at exceedingly
slow rates such that over very long periods of time, radioactive materials will escape to the ambient environment.
Transmission rates as low as 5×10^{-11} feet/sec calculated
for some shales, for example, might appear reassuring
inasmuch as the total movement in 1000 years would be roughly 1.5 feet. Unfortunately, such a simple calculation assumes that the transmission rate is applicable over long
periods of time, that the media will behave in accord with
the calculated model, that the system is isotropic, and that
unforseen changes will not occur in the intervening interval.
No simple geological analogs exist with comparable time
scales.

In reality the geological problem of containment of
wastes is much more complicated than the already complex
permeability problem alone implies (Schwartz, 1975). Another
desirable attribute of the geological formations would be the
ability to adsorb radioactive materials, an attribute which
might not be consistent with the lowest transmission rate
(Serne et al, 1977).

Transmission and absorption characteristics are, of

course, only a few among dozens of geological criteria which a suitable nuclear waste disposal site must meet (a host of economic and social criteria are not discussed here). Geological stability, more familiar from power plant siting controversies, is another. Will areas stable in the past remain stable in the future, and if so for how long? In geological parlance, is the past the key to the future? Salt may provide an excellent storage facility if it is assumed that retrieval will never take place (a notion of permanence rather at variance with customary human affairs). However, if preliminary experimental work at several laboratories is borne out (the structural strength of salt formations appears to be exceedingly low at the temperatures likely to be generated by the radioactive wastes and at the expected depths of burial) a repository in salt might tend to close up, requiring insertion of shafts to assure retrieval of wastes, reducing the integrity of the site.

Decisions regarding the location of sites for the disposal of long lived radioactive wastes, like all social decisions, do not rest solely upon information from the geological sciences or from any single body of knowledge. A too simple, but useful, illustration of the necessity of integrating social, economic, and political considerations in selecting site locations for waste disposal is suggested by the relationship between distance of transport from users of radioactive materials and availability of suitable geological formations for disposal. Shales, for example, might be intrinsically better but much farther from the locations of the users, a situation demanding a trade-off between risks of transport and risks in storage.

A recent report of the Nuclear Fuel Cycle Committee of the California Energy Resources Conservation and Development Commission (Varanini and Maullin, 1978, p. 9) states that "Since geology is not a predictive science, it may not be possible to reduce the uncertainties inherent in geologic disposal to a level where there is a high degree of confidence that the wastes can be confined for 100,000s of years." While the earth scientist may bristle at this sweeping conclusion, he may not wish to wholly disavow it. What is needed is a better articulation of the nature of the uncertainty of predictions about the earth and its behavior in relation to the alternative policy choices.

Some Conclusions

Every choice among alternative energy sources, uses, or technologies involves an alteration of the environment. Even increasing reduction of effluents to land, air, and

water through strengthened pollution control intended to
improve the environment often demands additional energy. It
is easier then to envision any number of diaster scenarios
associated with our utilization of different energy sources.
Similarly, one can point to a litany of failures to protect
the environment in the quest for energy, particularly cheap
energy. Aware of these risks and mindful of Pollyana, I
should like to suggest, nonetheless that ways can and will be
found to utilize diverse energy sources, that in the forsee-
able future we will have continuing social and technological
changes potentially both conserving and despoiling of the
environment, and that these will demand increasingly sophisticated knowledge about the earth and its environment. We
will always know less at the time of decision than we would
like to, it is only to be hoped that as time moves on, new
decisions will benefit from new knowledge. It is a paradox
that while we always wish to know more, we don't use the
knowledge we already have. Aside from the fact that we have
no choice but life on this earth, my optimism derives in part
from the observation of the resiliency of the earth over
periods of time. We may today be too quick to use the word
fragile in describing the environment. It implies both a
lack of resiliency and a stability which earth history and
processes belie. At the same time, the simple word "trade-
off" may be too glibly used if one is talking about global
CO_2, or synthetic materials created by man which have no
counterparts in nature.

The United States has passed through five to ten years
of a resurgence of what philosopher and historian of ideas,
George Boas, characterized in history as the kind of time
when the concept of Vox Populi was in the ascendancy. Boas
(1969, p.4) described this philosophy in the following terms:

"The People are assumed to have an infallible source
of knowledge, knowledge that is self-substantiated,
requiring no analysis or criticism. The proverb is
in this respect related to one of the many forms of
cultural primitivism, the form that maintains that
nature is better than art, that instinct is better
than learning, that feelings are wiser than reason,
that the "heart" is sounder than the "mind".

For our era it may well be that this turn from the mind
to the heart was essential to the maintenance of the people's
confidence in democratic processes and decision making, and
to the sustenance of a healthy skepticism about the infall-
ibility of "experts". Recent reports and polls in the press,
however, indicate that the public's confidence in scientists
remains high. In such an environment the earth sciences

should have an important role to play in illuminating choices and in resolving conflicts in the formulation of public policies which attempt to respond both to the demand for new and expanded sources of energy and for protection of the environment. The task of the earth scientist participating in the continuing debates about energy and environmental policy is not made easier by the mismatch between the long time scales of earth processes and the short time horizons of political decision makers. This discordance heightens rather than diminishes the need for such participation.

REFERENCES

Boas, George, 1969, Vox Populi: Essays in the History of an Idea, Johns Hopkins Press, Baltimore.

Bower, B.T., ed., 1977, Regional Residuals Environmental Quality Management Monitoring, Resources for the Future Research Paper R-7, 230 pp.

Cooke, R.U., and Reeves, R.W., 1976, Arroyos and Environmental Change in the American South-West, Oxford, Clarendon Press, 213 pp.

Delson, J.K., and Frankel, R.J., 1972, Residuals Management in the Coal-Energy Industry, Resources for the Future, manuscript.

Howe, C.W., 1977, A coordinated set of economic, hydro-salinity and air quality models of the Upper Colorado River Basin with applications to current problems, Chap. 4 in Bower, above, 1977, pp. 108-139.

Hubbert, M.K., 1949,"Energy From Fossil Fuels", Science, v. 109, pp. 103-109.

James, I.C., II, and Steele, T.D., 1977,"Application of Re-siduals Management for Assessing the Impacts of Alterna-tive Coal Development Plans on Regional Water Resources", 3rd International Hydrology Symp., Colo. State Univ., June 27-29.

Konikow, L.F., 1975,"Internatl. Conf. on Environmental Sensing and Assessment",v. 2, paper 20-3, 6 pp.

Likens, G.E., Bormann, F.H., and Johnson, N.M., 1972, Acid Rain, Environment, v. 14, pp. 33-40

Myers, D., Dorset, P., and Parker, T., 1977,"U.S. Envir. Protection Agency, Oil Shale and the Environment", EPA - 600/9-77-033, 29 pp.

National Academy of Sciences, 1977, Pond on Energy and Climate, 40 pp.

Ottar, B., 1975,"Organization of Long Range Air Pollution in Monitoring in Europe", Internatl. Conf. on Environmental Sensing and Assessment, v. 2, paper 32-6, 6 pp.

Schwartz, F.W., 1975,"On Radioactive Waste Management: An Analysis of the Parameters Controlling Subsurface Contaminant Transfer:, Jour. of Hydrol., v. 27, pp. 51-71.

Serne, R.J., Rai, D., Mason, M.J., and Molecke, M.A., 1977, Batch Kd measurements of nuclides to estimate migration potential at the proposed waste isolation pilot plant in New Mexico, Battelle Pacific Northwest Labs, PNL-2448, 47 pp.

U.S. Environmental Protection Agency, 1977,"Energy/ Environment II, Decision Series", EPA 600/9-77-012, 563 pp.

Varanini, E.E. III, and Maullin, R.L., 1978, Status of nuclear fuel reprocessing spent fuel storage and high-level waste disposal, Nuclear Fuel Cycle Comm. Calif. Energy Resources Conservation and Development Commission, 40 pp.

Wolman, M.G., 1971,"The Nation's Rivers", Science, v. 174, pp. 905-918.

Womack, W.R., and Schumm, S.A., 1977,"Terraces of Douglas Creek, Northwestern Colorade: An Example of Episodic Erosion", Geology, v. 5, pp. 72-76.

Index